I0493098

Learning to Fly

by Claude Grahame-White and Harry Harper

CONTENT

AUTHORS' NOTE

This book is written for the novice--and for the novice who is completely a novice. We have assumed, in writing it, that it will come into the hands of men who, having determined to enter this great and growing industry of aviation, and having decided wisely to learn to fly as their preliminary step, feel they would like to gain beforehand--before, that is to say, they take the plunge of selecting and joining a flying school--all that can be imparted non-technically, and in such a brief manual as this, not only as to the stages of tuition and the tests to be undergone, but also in regard to such general questions as, having once turned their thoughts towards flying, they take a sudden and a very active interest.

It has been our aim, bearing in mind this first and somewhat restless interest, to cover a wide rather than a restricted field; and this being so, and remembering also the limitations of space, we cannot pretend--and do not for a moment wish it to be assumed that we pretend--to cover exhaustively the various topics we discuss. Our endeavour, in the pages at our disposal, has not been to satisfy completely this first curiosity of the novice, but rather to stimulate and strengthen it, and guide it, so to say, on lines which will lead to a fuller and more detailed research.

It is from this point of view, as a short yet comprehensive introduction, and particularly as an aid to the beginner in his choice of a school, and in what may be called his mental preparation for the stages of his tuition, that we desire our book to be regarded.

C. G.-W. H. H.

April, 1916.

CHAPTER I

THEORIES OF TUITION

Only eight years ago, in 1908, it was declared impossible for one man to teach another to fly. Those few men who had risen from the ground in aeroplanes, notably the Wright brothers, were held to be endowed by nature in some very peculiar way; to be men who possessed some remarkable and hitherto unexplained sense of equilibrium. That these men would be able to take other men--ordinary members of the human race--and teach them in their turn to navigate the air, was a suggestion that was ridiculed. But Wilbur Wright, after a series of brilliant flights, began actually to instruct his first pupils; doing so with the same care and precision, and the same success, that had characterised all his pioneer work. And these first men who were taught to fly on strange machines--as apart from the pioneers who had taught themselves to fly with craft of their own construction--made progress which confounded the sceptics. They went in easy and leisurely fashion from stage to stage, and learned to become aviators without difficulty, and mainly without accident.

After this, increasing in numbers from two or three to a dozen, and from a dozen to fifty and then a hundred, the army of airmen grew until it could be totalled in thousands. Instead of being haphazard, the teaching of men to fly became a business. Flying schools were established; courses of tuition were arranged; certain pilots specialised in the work of instruction. It was shown beyond doubt that, instead of its being necessary for an aviator to be a species of acrobat, any average man could learn to fly.

Certainly a man who intends to fly should be constitutionally sound; this point is important. When in an aeroplane, one passes very quickly through the air, and such rapid movement--and also the effect of varying altitudes-- entail a certain physical strain. A man with a weak heart might find himself affected adversely by flying; while one whose lungs were not sound might find that his breathing was impeded seriously by a swift passage through the air. More than one fatality, doubtful as to its exact cause, has been attributed to the collapse of a pilot who was not organically sound, or who ascended when in poor health. And here again is an important point. No man, even a normally healthy man, should attempt to pilot a machine in flight when he is feeling unwell. In such cases the strain of flying, and the effect of the swift motion through the air, may cause a temporary collapse; and in the air, when a man is alone in a machine, any slight attack of faintness may be sufficient to bring about a fatality.

A fair judgment of speed, and an eye for distance, are very helpful to the man who would learn to fly, and it is here that a man who has motored a good deal, driving his own car, is at advantage at first over one who has not. But otherwise, and writing generally, any man of average quickness of movement, of average agility, can learn without difficulty to control an aeroplane in flight. It is wrong to imagine that exceptional men are required. An unusual facility, of course, marks the expert pilot; but we are writing of men who would attain an average skill.

There has been discussion as to the age at which a man should learn to fly, or as to the introduction of age limits generally in the piloting of aircraft. But this introduces a difficult question; one which depends so entirely on the individual, and regarding which we need the data that will be provided by further experience. Some men retain from year to year, and to a remarkable extent, the faculties that are necessary; others lose them rapidly. The late Mr. S. F. Cody was flying constantly, and with a very conspicuous skill, at an age when he might have been thought unfit. But then he was a man of a rare vitality and a great enthusiasm--a man who, though he flew so often, declared that each of his flights was an "adventure." Taking men in the average one may say this: the younger a man is, when he learns to fly, the better for him. Much depends, naturally, on the sort of flying he intends to do after he has attained proficiency. If he is going to fly in war, or under conditions that impose a heavy strain, then he must be a young man. But if he intends to fly for his own pleasure, and under favourable conditions, then this factor of age loses much of its importance, and it is only necessary that a man should retain say, an ordinary activity, and a normal quickness of vision and of judgment.

Flying is not difficult. It is in a sense too easy, and this is just where its hidden danger lies. If a pupil is carefully taught, and flies at first only when the weather conditions are suitable, he will find it surprisingly easy to pilot an aeroplane. That it is not dangerous to learn to fly is proved daily. Though hundreds and thousands of pupils have now passed through the schools, anything in the nature of a serious accident is very rarely chronicled. This immunity from accident is due largely to the care and experience of instructors, and also to the fact that all pupils pass through a very carefully graduated tuition, and that no hazardous flights are allowed; while another and an important element of safety lies in the fact that no flying is permitted at the schools unless weather conditions are favourable. It is now a fair contention that, provided a man exercises judgment, and ascends only in weather that is reasonably suitable, there is no more danger in flying an aeroplane than in driving a motor-car.

Much depends of course on the dexterity of the pupil, and particularly on his manual dexterity--on what is known, colloquially, as "hands." Some men, even after they have been carefully taught, are apt to remain heavy and clumsy in their control. Others, though, seem to acquire the right touch almost by instinct; and these are the men who have in them the making of good pilots. Horsemen refer to "hands" when they speak of a man who rides well; and in flying, if a man is to handle a machine skilfully, there is need for that same instinctive delicacy of touch.

Nowadays, when a pupil joins a well-established flying school, he finds that everything is made easy and pleasant for him. Most men enjoy very thoroughly the period of their tuition. A friendly regard springs up between the pupils and their instructors, and men who have learned to fly, and are now expert pilots, bear with them very pleasant reminiscences of their "school" days. But there were times, and it seems already in the dim and distant past, when learning to fly was a strange, haphazard, and hardly pleasant experience; though it had a sporting interest certainly, and offered such prospects of adventure as commended it to bold spirits who were prepared for hardship, and had a well-filled purse. The last requirement was very necessary. In the bad old days, amusing days though they were without doubt, no fixed charge was made to cover such breakages, or damage to an aeroplane, as a pupil might be guilty of during his period of instruction. These items of damage--broken propellers, planes, or landing gear--were all entered up very carefully on special bills, and presented from time to time to the dismayed novice; and a man who was clumsy or impetuous found learning to fly an expensive affair. There was a pupil who joined a school soon after Bleriot's crossing of the Channel by air. It was a monoplane school; and the monoplane, unless a man is careful and very patient, is not an easy machine to learn to fly. This beginner was not patient; he was indeed more than usually impetuous. His landings, in particular, were often abrupt. He broke propellers, frequently, to say nothing of wings and of alighting gear. And of all these breakages a note was made. Bills were handed to him--long and intricate bills, with each item amounting to so many hundreds of francs. Having a sense of humour, the pupil began to paper his shed with these formidable bills, allowing them to hang in festoons around the walls. What it cost him to learn to fly nobody except himself knew. He paid away certainly, in his bills for breakages, enough money to buy several aeroplanes.

This was in the early days, when aviators were few and all flying schools experimental. To-day a pupil need not concern himself, even if he does damage a machine. Before beginning his tuition he pays his fee, one definite sum which covers all contingencies that may arise. It includes any and all damage that he may do to the aircraft of his instructors; it covers also any third-party claims that may be made against him--claims that is to say from any third person who might be injured in an accident for which he was responsible. This inclusive fee varies, in schools of repute, from ?5 to ?00.

The modern aerodromes, or schools of flight, at which a pupil receives his tuition, have been evolved rapidly from the humblest of beginnings. The first flying grounds were, as a rule, nothing more than open tracts of land, such as offered a fairly smooth landing-place and an absence of dangerous wind-gusts. Then, as aviation developed, pilots came together at these grounds, and sheds were built to house their craft. And after this, quickly as a rule, an organisation was built up. Beginning from rough shelters, erected hastily on the brink of a stretch of open land, there grew row upon row of neatly-built sheds, with workshops near them in which aircraft could be constructed or repaired. And from this stage, not content with the provision made for them by nature, those in control of the aerodromes began to dig up trees, fill in ditches and hollows, and smooth away rough contours of the land, so as to obtain a huge, smooth expanse on which aircraft might alight and manoeuvre without accident. And after this came the building up of fences and entrance gates, the erection of executive offices and restaurants, the provision of telephone exchanges and other facilities--the creation in fact of a modern aerodrome.

A pupil to-day, if he decides to learn to fly, finds he has an ample choice in the matter of a school. He may feel indeed that there is almost an embarrassment of facilities. But there are certain very definite requirements, in regard to any modern flying school; and if a novice bears these in mind, and thinks of them carefully when he is considering what school he shall join, he cannot go far wrong. First there is the question of the aerodrome on which, and above which, the pupil will undergo his instruction. This should be of ample size and of an adequately smooth surface; and it should be so situated, also, that it is free from wind eddies and gusts, such as are set up by hills, woods, or contours of the land, and are likely to inconvenience a novice when he makes his first flights. The best position for an aerodrome is in a valley, not abrupt but gently sloping. With a flying ground so placed, shielded well by nature on every hand, it may prove sufficiently calm for instruction even on days when there is a gusty wind blowing across more exposed points; and such a natural advantage is of importance for a pupil. It may mean that he is obtaining his tuition from day to day, when other pupils, learning to fly at grounds less favourably situated, have to remain compulsorily idle, waiting either for the wind to drop, or to veer to some quarter from which their aerodrome is sheltered.

It is very necessary, of course, in the operation of a flying school, that there should be competent instructors; also a sufficient number of these to prevent them from being over-taxed, or having more pupils at any one time than they can handle conveniently. And it is greatly to the advantage of a pupil if these instructors have been chosen with an intelligent care. A man may be a capable pilot, and yet not have the temperament that will suit him for imparting his knowledge to others. The instructor who, besides being a fine flyer, has the patience and sympathy of a born teacher, is by no means easy to find. A school which does find such men, and retains their services, offers attractions for a pupil which--in any preliminary visit he pays to a school before joining it--he should look for keenly. And he should make certain, too, that the school has a staff of skilled and experienced mechanics.

Another indispensable feature of a school is a sufficient number of aeroplanes, machines suited specially for the purposes of tuition, and maintained at a high efficiency. It has been no uncommon thing--though here again one is writing of the past--for the total resources of a school to comprise, say, two machines. Hence a couple of smashes would put such a school temporarily out of action, and leave the pupils with nothing to do but kick their heels, and wait until the machines had been repaired. It is certainly an advantage, from the pupil's point of view, if there are well-equipped workshops in connection with the school he joins; also if the proprietors of his school have an ample supply of engines. With facilities for repair work immediately at hand, and with a spare engine ready at once to put in a machine--while one that has been giving trouble is dealt with in the engine-shop--there should always be a full complement of craft for the work of instruction. When workshops are in operation in connection with a school an opportunity is usually provided, also, for a novice to gain some knowledge as to the mechanism and working of the aero-motor: and this of course will be useful to him.

There has been discussion as to the type of aeroplane on which one should learn to fly; but in this question, as in that of an age limit for airmen, it is extremely difficult, besides being unwise, to attempt to frame a hard-and-fast rule. The monoplane, for instance, is not an easy machine to learn to fly: it is not easy, that is to say, compared with certain types of biplane. Yet numbers of pupils have been taught on monoplanes, and this without accident. There is also a question whether, among biplanes, it is best to learn on a tractor machine--one that is to say with the engine in front of the main planes--or on a "pusher" type of craft; this last mentioned having its motor behind the planes. Aeroplanes of both types are in use; and it would be advantageous, of course, for a novice to accustom himself to handle either. But from the point of view of those who operate large flying schools, and have to weigh one point against another, and eliminate so far as possible the elements of risk or difficulty, there are very distinct advantages in a "pusher" biplane, such as is illustrated facing page 34. The control of such a machine is simple, and can be grasped quite readily. It provides the novice, when he is seated in it, with a clear and unobstructed view of the ground immediately in front of and below him; and this, in the early stages of tuition, is an extremely important point. A craft of such a type, also, when built specially for instruction, can be given a very strong alighting gear, and this makes for safety when a pupil is in his first tests, and may be guilty of an abrupt or rough descent. Again, while such a school machine as this is engined adequately, it is at the same time comparatively slow in flight, and has the advantage also that it will alight at slow speeds. In the air, too, it has a large measure of stability, and is not too rapid in its response to its controls. It gives a pupil what is very necessary for him in his first flights, and that is a certain latitude for error. It is safe to say, indeed, without being dogmatic, that a "pusher" biplane of the type illustrated, if constructed specially for school work, offers a pupil two very clearly marked advantages. These are: (1) A craft which he can learn to fly quickly; and (2) A machine on which he can pass through his tuition with the least risk of accident.

This last-mentioned point is, naturally, one of extreme importance. It is very necessary, apart from any question of personal injury, that a pupil should be protected during his tuition from anything in the nature of a bad smash. A man should start to learn to fly with full confidence; the more he has the better, provided it is tempered with caution. And if he can go through his training without accident, and preserve the steadily growing confidence that his proficiency will give him, he is on the high road to success as a pilot. But if he meets with an accident while he is learning--some sudden and quite unexpected fall--this may have a serious and a permanent influence on his nerves, even if he escapes without injury. It happened frequently in the early days that a promising pupil, a man who showed both confidence and skill, had his nerve ruined, and all his "dash" taken from him, by some unlucky accident while he was learning to fly.

There are certain minor points a pupil should consider when he selects a flying school--points which have reference mainly to his own comfort and convenience. He will prefer, for instance, other things being equal, a school that is near some large town or city, and not buried away inaccessibly. It is a convenience also, and one that facilitates instruction, if a pupil can obtain, quite near the aerodrome, rooms where he can live temporarily while undergoing his instruction, and so be able to reach the flying ground in a minute or so, whenever and at any time the weather conditions are favourable. It is a convenience again if, either on the aerodrome itself or immediately adjacent, there is a canteen or restaurant where meals and other refreshments can be obtained. Dressing-rooms and reading rooms, when provided by the proprietors of a school, add to the comfort of the novice while he is in attendance on the aerodrome. In winter, particularly, such facilities are required.

At a modern school, if it is well conducted, all heroics or exceptional feats are discouraged. Pupils who want to do wild things must be sternly repressed, even if only for the common good. The aim is to train a certain number of pupils, not hastening over the tuition but giving each man his full and complete course, and to do this with a minimum of risk. In the early days of flying there were remarkable exploits at the schools, and some very dangerous ones also. But nowadays the reckless, happy-go-lucky spirit has gone. Tuition is based on experience. Each pupil must submit to the routine, and listen attentively to the instructions given him. There are no short cuts--not at any rate with safety--in the art of learning to fly.

The question is asked, often, how long it should take a man to learn to fly. It is almost impossible, though, to specify any fixed time. A very great deal must depend on the weather. A pupil who joins a school in the summer is more likely, naturally, to complete his tuition quickly than one who begins in the winter. In periods when there are high and gusty winds it may be necessary to suspend school work for several days. But at such times the pupil need not be completely idle. Lectures on aviation are organised sometimes by the schools; while a pupil should have opportunities also--as has been mentioned before--of going into the engine-shop and studying the repair and overhaul of motors and machines.

It is on record that a pupil has learned to fly in a day, even in a few hours; but here the circumstances, and the men, were exceptional. Such an unusual facility represents one extreme; while as another, it may happen that a man, owing to a combination of adverse circumstances, is six months before he gains his certificate of proficiency. It may be taken, as a rule, that a pupil should set aside say a couple of months in order to undergo thoroughly, and without any haste, his full period of tuition. School records prove, as a rule, that the pilots who learn to fly abnormally quickly are apt to experience an abnormal number of accidents at a later date, due principally to a lack of real sound knowledge, which they should have gained during the period of their tuition. One must learn to walk before one can run, and this takes time; and the remark applies aptly to aviation. It is very necessary for the pupil to spend as much time as he can on the aerodrome. Much is to be learned, by an observant man, apart from the actual time during which he is engaged with his instructor. If he watches men who are highly skilled, he may gain many useful hints, though he himself is on the ground.

CHAPTER II

TEMPERAMENT AND THE AIRMAN

As aviation passed from its earliest infancy, and a number of men began to fly, the temperament of the individual pupil, and the effect of this temperament on his progress as an aviator, began to reveal itself. And temperament does play a large part in flying; as it does in any sport in which a man is given control of a highly sensitive apparatus, errors of judgment in the handling of which may lead to disaster. It is not, as a rule, until he has passed through his early stages of tuition, and has begun to handle an aeroplane alone, and is beyond the direct control of his instructor, that the temperament of a pupil really plays its part. Up to this point he is one among many, conforming to certain rules, and obliged to mould himself to the routine of the school. But when he begins to fly by himself, and particularly when he has passed his tests for proficiency, and is embarking, say, on cross-country flights, then this question of temperament begins really to affect his flying.

All men who learn to fly--numbering as they do thousands nowadays--cannot be endowed specially by nature for their task. There is indeed a wide latitude for temperamental differences--always provided that nothing more is required of a man than a certain average of skill. But if a man is to become a first-class pilot, one distinctly above the average, then the question of his temperament, as it influences his flying, is certainly important.

A rough classification of the pupils at a school--just a preliminary sorting of types--shows as a rule the existence of two clearly-marked temperaments. One is that of the man who is deliberate, whose temperament guards him from doing anything perfunctorily or in a hurry; the other is that of a man--a type frequently encountered nowadays--who while being quick, keen, and intelligent, mars these good qualities by a temperamental impatience which he finds it difficult or impossible to control, and which makes him irritable and restless at any suggestion of delay.

Now the first of these men need not to be wholly commended, nor the second entirely condemned. A capacity for deliberation, both in study and in practice, is very useful when learning to fly. It will protect a man from many errors, and render his progress sure, though it may be slow. But something more than deliberation is required in the aviator of distinction. There must be the vital spark of enterprise, the temperamental quality which is known as "dash," the quick action of the mind, in difficulty or peril, that will carry certain men to safety through many dangers. This imaginative power is possessed as a rule, though in ways that differ considerably, by the second type of pupil we have described--the restless, impatient man. But in his case this quality is, more often than not, marred by his instability; by the lack of that judgment which is so necessary to counterbalance imagination, but which is, unfortunately, not so often found.

A man who decides to become an aviator, and particularly if he intends to fly professionally, should ask himself quite seriously if his temperament is likely to aid him, or whether perhaps it may not be a danger. This point is certainly one of importance, though it cannot be stated directly or decided in so many words. There is a vital question at least that the novice should ask himself; and this is whether his temperament, whatever its general tendency may be, includes a sufficient leavening of caution. In the navigation of the air caution is indispensable. A pupil must remind himself constantly that, though it appears easy--and is indeed easy--to learn to handle a machine in flight, no liberties must under any circumstances be taken with the air. Every instant a man is flying he needs to remember the value of caution. In the air one cannot afford to make mistakes.

Naturally there is an ideal temperament for flying; but it is one which, owing to the combination of qualities that are required, is very rarely met with. The man who possesses it is gifted with courage, ambition, "dash," and with a readiness in an emergency that amounts to intuition. And yet these positive qualities are, in the ideal temperament, allied to, and tempered by, a strong vein of prudence and of caution. The pilot has absolute system, method, and thoroughness in everything he does. The average pupil cannot hope to be so luckily endowed. But he can study his personality, and seek to repress traits that may seem harmful.

There is need in flying for a sound judgment, one that will enable a man to come to a decision quickly and yet accurately. Things happen rapidly in the air. It is one of the grim aspects of flying that, just at a moment when everything appears secure, a sudden disaster may threaten. So it is of vast importance to a pilot, if he has to fly regularly, that he should have an instinctive and dependable judgment; a capacity for deciding quickly and without panic; a capacity, when several ways present themselves of extricating himself from some quandary, of being able to choose the right one, and of not having to think long before doing so. This implies a combination really of judgment and resource. The man of confidence, the man of resource, is well endowed for flying. But he must not be over-confident. The over-confident man is a menace to himself and to others. It is not a proper spirit at all in which to approach aviation. We do not know enough about the navigation of the air to be in the least over-confident. The spirit, rather, should be one of humility--a determination to proceed warily, and to make very certain of what limited knowledge we do possess.

Two of the worst traits in an aviator are impatience and irritability. A man who has these temperamental drawbacks in a form which is strongly marked, and who cannot control them, should not think of becoming an aviator. The man who is impatient and irritable finds himself out of harmony with the whole theory of aerial navigation. There is a long list of "don'ts" in flying; in the handling of one's machine, in the weather one flies in, in all the feats that one should attempt and leave alone. A number of details must be memorised, and must never be forgotten or overlooked, trivial though some of them may seem. The frame of mind of the man who flies must be alert, yet quiet and reposeful; he must be clear-headed, not hot-headed. The man who is in a hurry, who ignores details when he sets out on a flight, is the man who runs risks and is bound sooner or later to pay the penalty. The perils of recklessness in flying are very great. The man who "takes chances," who thinks he can do something when, as a matter of fact, he has neither sufficient knowledge or experience, runs a very grave and constant risk. It is the thoughtful, considering frame of mind, particularly in a pupil, which is the safe one; but this must not be taken to imply a type of man who lacks power of action. Initiative, and a quick capacity for action, are most necessary in aviation. New problems are being faced continually, and the brain succeeds which is the most active and original.

CHAPTER III

FIRST EXPERIENCES WITH AN AEROPLANE

(AS DESCRIBED BY MR. GRAHAME-WHITE)

After a period of ballooning, which offers experience for an aviator in the judging of heights and distances, and in growing accustomed to the sensation of being in the air, I devoted a good deal of time and attention--more indeed at the time, and in view of my other responsibilities, than I could reasonably spare--to a study of the theory of aeroplane construction, and to the making of models. This was prior to 1909; Bleriot had not yet flown the Channel in his monoplane. But when he did I put models aside, and determined to buy an aeroplane and learn to fly.

At the end of August, 1909, so that I might inspect the various aeroplanes that were then available, and they were few enough, I went to Rheims, in France, and attended the first flying meeting the world had seen. At the aerodrome I met and talked with the great pioneers: with Bleriot, fresh from his cross-Channel triumph; with Levavasseur, the designer of the beautiful but ill-fated Antoinette monoplane, which had, through engine failure, let Hubert Latham twice into the Channel during his attempts to make the crossing; with Henry Farman who, fitting one of the first Gnome motors to a biplane of his own construction, flew for more than three hours at Rheims, and created a world's record; and also with M. Voisin, whose biplane was then being flown by a number of pilots.

Finally, after careful consideration, I made a contract with M. Bleriot to purchase from him, at the end of the meeting, a monoplane of a type that appeared first at Rheims, and of which there was not another model then in existence. This machine differed considerably from the one with which M. Bleriot had flown the Channel. His cross-Channel monoplane was a single-seated craft fitted with an air-cooled motor of about 25 h.p. The machine I agreed to buy at Rheims, and which was known as Bleriot No. XII., would carry two people, pilot and passenger, while it had an 8-cylinder water-cooled motor developing 60 h.p.--an exceptional power in those days. The position of the occupants, as they sat in the machine, differed from the arrangement in the cross-Channel Bleriot. In the latter the pilot sat in a hull placed between the planes, and with his head and shoulders above them. But in this new and larger machine the pilot and passenger sat in seats which were placed below the planes.

The craft was, as a matter of fact, an experiment, being built almost purely for speed; hence its powerful motor. M. Bleriot's idea, in constructing it, was to have a machine with which he might win the Gordon-Bennett international speed race at Rheims. But this hope he did not realise; nor did I obtain delivery of the craft I desired. Bleriot, flying alone in this big monoplane, started in a speed flight for the Gordon-Bennett; but he was only a quarter of the way round the course, on his second lap, when the machine was seen to break suddenly into flames and crash to the ground from a height of 100 feet. It was wrecked entirely, but Bleriot was fortunate enough to escape with nothing worse than burns about the face and hands, and a general shock. The cause of the accident was that an indiarubber tube, fixed temporarily to carry petrol from the tank to the carburettor, had been eaten through and had permitted petrol to leak out, and to ignite, on the hot exhaust pipes of the motor.

The destruction of this monoplane was, to me, a great disappointment. No other machine of the type was in existence, and I learned that it would take three months to build one. M. Bleriot promised, however, to put a machine in hand at once; and, as a special concession, I obtained permission to go daily to the Bleriot factory and superintend the construction of my own machine. This I did for a full period of three months, working daily from 6 a.m. to 6 p.m., and gaining some valuable knowledge as to aeroplane construction.

On November 6, 1909, after delays which had tried my patience sorely, I obtained delivery of the new machine--a replica of the craft that had been destroyed at Rheims. It was too late that day to begin any trials, so I and a friend who was with me arranged with M. Bleriot's mechanics that we would be at Issy-les-Moulineaux early next morning, and there put the craft through its preliminary tests. I can remember we went to bed early, but sleep was impossible; we were both too excited at the prospect that lay before us. So presently we got up--this was at 2 a.m.--and drove out to the flying ground.

It was pitch dark when we arrived at the aerodrome, but the morning promised to be favourable. Foggy it was; but there was no wind, and the fog seemed likely to clear. We roused the caretaker, and, after lengthy explanations and considerable monetary persuasion, induced him to open the shed and allow us to prepare the machine for its first flight. Then we waited for the mechanics and the first rays of dawn. We felt a desire to get the big engine started up, but had been warned of the risk of doing this without the help of mechanics. Time passed and still the mechanics did not come. At last, there being now sufficient light, we tied the aeroplane with ropes to a fence, so as to prevent its leaping forward, and then started up the motor by ourselves. I swung the nine-foot propeller--the only way of starting the engine; and at the first quarter-turn the motor began to fire. Then, as is quite usual, there was an incident that had been unforeseen in our excitement. We had forgotten to take up the slack of the rope; and the consequence was that, as the engine started, the machine gave a bound forward that was sufficient to knock me down. But I was unhurt, and picked myself up quickly. Then I hurried round to the driving seat and took my place at the control levers, motioning to my friend, who was looking after the ropes, to cast these loose and jump into the seat beside me. This was easier said than done. Directly he released the ropes the machine began to move across the ground, gathering speed very quickly; but he managed somehow, before the machine was running too fast, to scramble into the seat beside me.

Off we started across the aerodrome, the monoplane gaining a speed of 40 or 50 miles an hour. I did not attempt to rise from the ground, feeling it very necessary at first to grow familiar with the controls. So we sped along the ground for a distance of about a mile. Then, on nearing the far end, I slowed down the motor and our speed dropped to about 20 miles an hour. I wanted to turn the machine round on the ground and run back again towards our starting point. But such a manoeuvre, particularly for the novice, is far from easy. As the speed of the machine is reduced, the pressure of air on the rudder is lessened and so it loses its efficiency--in the same way that a ship is difficult to steer when she begins to lose way. We were faced also by another and a graver difficulty. Confused by the fog, which still hung over the aerodrome, I had misjudged our position. We found we were much nearer the end of the ground than I had imagined. In front of us there loomed suddenly a boundary wall, against which it seemed probable we should dash ourselves. There were no brakes on the machine; no way of checking it from the driving seat. Our position seemed critical.

It was now that I shouted to my friend, telling him to jump out of the machine as best he could, and catch hold of the wooden framework behind the planes, allowing the machine to drag him along the ground, and so using the weight of his body as a brake. This, with great dexterity, he managed to do, and we came to a standstill not more than a foot or so from the wall. This proved a chastening experience; we pictured our aeroplane dashed against the wall, and reduced to a mass of wreckage. Very cautiously we lifted round the tail of the machine. It was impossible to switch off the motor and have a rest, because, if we had stopped it, we should not have been able to start it again without our gear, which was away on the other side of the ground.

Now, having got the machine into position for a return trip across the aerodrome, I accelerated the engine, and we started off back. For about twenty minutes, without further incident, we ran to and fro; and now I felt that I had the machine well in control--on the ground at any rate. And so the next thing was to rise from the ground into the air. I told my friend my intention, calling to him above the noise of the motor; and I admired him for the calm way in which he received my news. I should not have been surprised if he had demanded that I should slow up the machine and let him scramble out. In those days it was thought dangerous to go up even with a skilled and more or less experienced pilot. How much greater, therefore, must have seemed the risk of making a trial flight with me--a complete novice in the control of a machine. But my friend nodded and sat still in his seat. So I accelerated the motor and raised very slightly our rear elevating plane. And then we felt we were off the ground! There was no longer any sensation of our contact with the earth--no jolting, no vibration. In a moment or so, it seemed, the monoplane was passing through the air at a height of about 30 feet. This, to our inexperienced eyes, appeared a very great altitude; and I made up my mind at once to descend. This manoeuvre, that of making contact with the ground after a flight, I had been told was the most difficult of all. It is not surprising that this should be so. Our speed through the air was, at the moment, about 50 miles an hour; and to bring a machine to the ground when it is moving so fast, without a violent shock or jar, is a manoeuvre needing considerable judgment. But, remembering that the main thing was to handle the control lever gently, I managed to get back again to the aerodrome without accident; and after this we turned the machine round again and made another flight.

The fog had cleared by now, and we were surprised to see a number of people running across the ground towards us. First there came the tardy mechanics; and with them were a number of reporters and photographers representing the Paris newspapers. These latter had--though I only found this out afterwards--been brought by the mechanics in the expectation of being able to record, with their notebooks and cameras, some catastrophe in which we were expected to play the leading parts. Knowing the powerful type of monoplane I had acquired, a machine not suited for a novice, the mechanics had felt sure some disaster would overtake me. But, as it happened, their anticipations were not fulfilled. The journalists and photographers did not, however, have a fruitless journey. Though there was nothing gruesome to chronicle, they found ample material, when they learned of them, in the early morning adventures of myself and my friend with this 60 h.p. monoplane. Next day, in fact, our exploits were given prominence in the newspapers, and I received a number of congratulatory telegrams; not forgetting one of a slightly different character which came from M. Bleriot. He was flying at the time in Vienna, and he warned me of the dangers of such boldness as I had displayed--having regard to the speed and power of my machine--and pleaded with me for a greater caution.

CHAPTER IV

THE CONTROLLING OF LATEST-TYPE CRAFT

People are puzzled, often, when they try to explain to themselves how it is that an aeroplane, which is so much heavier than air, manages to leave the ground and to soar in flight. When balloons or airships ascend, it is realised of course that the gas, imprisoned within their envelopes, draws them upward. But the aeroplane--weighing with pilot, passenger, and fuel perhaps several thousand pounds--rises without the aid of a gas-bag and with nothing to sustain it but narrow planes; and these do not beat, like the wings of a bird, but are fixed rigidly on either side of its body. How is the weight of machine and man borne through this element we cannot see, and which appears intangible?

The secret is speed--the sheer pace at which an aeroplane passes through the air. As a craft stands on the ground, its planes are inoperative. Power lies dormant in the air, but only when it is in motion, or when some object or apparatus is propelled through it at high speed. Have you stood on a height, in a gale, and felt an air wave strike powerfully against your body? The blow is invisible; but you yield a step, gasping; and, had you wings at such a moment, you would not doubt the power of the wind to sweep you upward. This is the force the aeroplane utilises.

If, on a calm day, you accelerate your motor-car to 60 miles an hour, the air sweeps past you in a powerful stream; just as it would if you were standing still, and there was a gale of wind. Instead of the wind possessing the speed, in this instance, it is you who provide it. The motor of an aeroplane, driving the propeller of the machine, turns this at 1000 or more revolutions a minute, and causes its curved blades to screw forward through the air as they turn, like those of a ship's propeller through water--or a gimlet into wood. The propeller, as it bores its way into the air, draws or pushes the aeroplane across the ground; and the speed grows rapidly until the air, sweeping with an increasing pressure beneath the planes, becomes sufficient to bear the craft in flight.

But the wing of an aeroplane would not sustain its load unless designed specially to act upon the air. A man, if he is unlucky enough to fall from a tall building, passes through the air at a high speed. His body obtains no support from the air; so he crashes to the ground. This is because his body is heavy, and presents only a small surface to the air. To secure a lifting influence from the air, it must be struck swiftly with a large, light surface.

Men go to Nature when building wings for aeroplanes, and imitate the birds. The wing of a bird arches upward from front to back, most of the curve occurring near the forward edge; and this shape, when applied to an aeroplane wing, is known as its camber. With an aeroplane wing, if its curve is adjusted precisely, the air not only thrusts up from below as a machine passes through it, but has a lifting influence also from above; an effect that is secured by the downward slope of the plane towards its rear edge. The air, sweeping above the raised front section of the plane, is deflected upward, and with such force that it cannot descend again immediately and follow the downward curve of the surface. So, between this swiftly-moving air stream, and the slope to the rear of the plane, a partial vacuum is formed, and this sucks powerfully upward. With a single wing, therefore, it is possible to gain a double lifting influence--one above and one below.

The building of aeroplanes, once their wing lift is known, becomes a matter of precision. According to the speed at which they fly, and the size and curve of their planes, machines will sustain varying loads. In some machines, as a general illustration--craft which fly fast--the planes may bear a load equal to 10 lbs. per square foot. In others the loading may be less than 3 lbs. per square foot.

Apart from raising a craft into the air, by the lifting power of its wings, there is the problem of controlling it when in flight. The air is treacherous, quickly moving. Gusts of abnormal strength, sweeping up as they do invisibly, may threaten to overturn a machine and dash it to earth. Eddies are formed between layers of warm and cold air. There are, as a craft flies, constant increases or lessenings of pressure in the air-stream that is sweeping under and over its wings; and all these fluctuations influence its equilibrium. Unless, therefore, a machine is automatically stable--and with craft of this type we shall deal later--the pilot must be ready, by a movement of the surfaces which govern the flight of the machine, to counteract quickly, with a suitable action of his levers, the overturning influence that may be exercised by a gust of wind. Here lies the art of flying. A man is given a machine which, by the action of its motor and propeller, will raise itself into the air; and it is his task, when the craft is once aloft, to manipulate it accurately and without accident, and to bring it to earth safely after he has made a flight.

In the description of controlling movements which follows we shall, for the sake of convenience, and for the sake also of brevity, deal only with the type of "pusher" biplane to which reference has been made already, and on which large numbers of pupils have been, and are being, trained to fly. This casts no aspersion whatever on tractor machines or on monoplanes. On either, if he has an inclination, a pupil can undergo his instruction, and do so usually with success. But explanation is rendered more easy, and there is less likelihood of a dispersal of interest, if one machine is selected for illustration; and our reasons for the choice of a "pusher" biplane, regarded from the point of view of tuition, have been explained already.

First, therefore, one may deal with raising the craft into the air, and causing it to descend. In the photograph of the school machine shown facing this page, it will be seen that the control surfaces are indicated by lettering. In front of the biplane, on outriggers, is the plane "A." This surface (aided in its action by a rear plane) governs the rise or descent of the machine. When the motor is started, and the propeller drives the biplane across the ground on its chassis B, the machine would, if this lifting plane was held in a negative position, continue to move forward on the earth and would make no attempt to rise. In order to leave the ground, when the speed of the machine is sufficient for its main-planes (C.C.) to become operative, and bear its weight through the air, the pilot draws back slightly towards him a lever, which is placed just to the right of his driving-seat and is held with the right hand. A photograph which shows this lever, and the other controls, appears facing page 36, the lever to which we are referring being indicated by the figure 1. The effect on the aircraft when the pilot draws back this lever--the motion being slight and made gently--is to tilt up the elevating plane A, and this in its turn, owing to the pressure of air upon it, raises the front of the machine. The result of this alteration in the angle of the craft is that it presents its main-planes at a steeper angle to the air. Their lifting influence is increased, with the result that--at an angle governed by the pilot with his movement of the elevating plane--they bear the machine from the ground into the air.

A reverse movement of the elevator reduces the lift of the main-planes; hence, when an aviator wishes to descend, he tilts down his elevator, bringing his machine at such an angle that it is inclined towards the ground. Then, switching off his engine so as to moderate the speed of his descent, and by such manipulations as may be necessary of his elevator, he pilots his craft to earth in a vol-plan? during which gravity takes the place of his motor, and he is able--by steadying his machine and bringing it into a horizontal position just at the right moment--to make a gentle contact with the ground.

A pilot must be able to do more than cause his aeroplane to ascend and to alight: he must have means to check the lateral movements which, under the influence of wind gusts, may develop while the biplane is in flight. At the rear extremities of the main-planes as illustrated in the photograph facing page 34--and marked D.D.--are flaps, or ailerons, which are hinged so that they may be either raised or lowered. These ailerons are operated, through the medium of wires, by the same hand-lever which governs the movement of the elevator. This lever is mounted on a universal joint, and can be moved from side to side as well as to and fro. Should the biplane tilt, while flying, say towards the left, the pilot moves his hand-lever sideways towards the right. This is a natural movement, the instinct being to move the lever away from the direction in which the machine is heeling. This movement of the lever has the effect of drawing down the ailerons on the left-hand side of the machine; on the side, that is to say, which is tilted down; and the depression of these auxiliary surfaces, increasing suddenly as they do the lifting influence of the main-planes to which they are attached, tend to thrust up the down-tilted wings, and so restore the equilibrium of the machine.

In the operation of his ailerons, combined with the use of his elevator, a pilot is given means to balance his craft while in flight. One should not gain the impression that an aeroplane is threatening ceaselessly to heel this way and that. This is not so. The machine has a large measure of stability, apart from any manipulation of its controls, and needs balancing only when some disturbance of the atmosphere affects its equilibrium. Under favourable conditions, such as a pupil will experience in his first flights, nothing more is necessary with the hand-lever than a very slight but fairly constant action; a similar motion, in a way, as is made by the driver of a motor-car when he maintains, by his "feel" on the wheel, his sense of control over the machine. In the controlling actions of an aeroplane--and this is a fact which tends sometimes to the confusion of the novice--nothing more is required, normally, than the most delicate of movements. The difference say between ascending, and skimming along the ground, is represented by a movement of the hand-lever of only a few inches. Delicate, sure, quick, and firm; such is the touch needed with an aeroplane.

With the one hand-lever, as we have shown, it is possible for a pilot to control the rise and descent, and also the lateral movements of his machine; and there remains only the steering to be effected--the movement from side to side, from right to left, or vice-versa. At the rear of the biplane, as shown facing page 34, will be seen two vertical planes, E.E. These, being hinged, will swing from side to side; and they exercise a sufficient influence, when working in the strong current of air that blows upon them when a machine is in flight, to steer it accurately in any direction. The pilot, to operate this rudder, rests his feet on a conveniently-placed bar, which is mounted on a central swivel, and allows the bar to be swung by a pressure of either foot.

When the pilot needs to make a turn say to the left, as he is flying, he presses his left foot forward. This swings the bar in same direction; and, by a simple connection of wires running to the tail of the machine, the rudders are made to swing over to the left also, and the machine turns in response to them. A similar movement to the right produces a right-hand turn. This foot rudder bar, being numbered 2, is shown in the picture facing page 36.

Apart from the movements we have described, which are extremely simple, a pilot needs also to maintain control over his motor. Near his left hand, fixed to the framework just at one side of his seat, are levers which govern the speed of the engine, also the petrol supply; while close to them is the switch by which the ignition can be switched on or off.

A final word is necessary here, perhaps, and it is this: the glamour and mystery which, in the early days, clung to the handling of an aeroplane has now been dispelled almost entirely. A well-constructed machine, flying under favourable conditions, requires surprisingly little control; what it does, one may almost say, is to fly itself.

CHAPTER V

THE STAGES OF TUITION

Flying schools--those which really can be described as such--have been in operation now for seven years; and during this time, with thousands of pupils going through their period of tuition, many very valuable lessons have naturally been learned. To-day, at a well-managed school, each stage in a pupil's instruction, mapped out as a result of experience, is arranged methodically and with care; the idea being that the novice should pass from one stage to another by a smoothly-graduated scale, facilitating his progress and reducing elements of risk.

It is in the early morning, and again in the evening, that the flying schools are most busy as a rule. At such times--morning and evening--the wind blows with least violence; and it is very necessary that a pupil, when he is handling craft for the first time, should have weather conditions which are favourable. Summer and winter, as soon as it is light, and granted conditions appear suitable, mechanics wheel the aeroplanes from the sheds, and the instructors begin their work. Should there be any doubt as to the weather, or as to the existence, say, of difficult air currents, an instructor will fly first, circling above the aerodrome at various heights, and satisfying himself, by the behaviour of his machine, whether it will be safe for the novices to ascend. If he pronounces "all well," school work begins in earnest, and continues--provided the weather remains favourable--until all the pupils have had a spell of instruction. Towards the middle of the day, and in the afternoon, it is quite likely the wind may blow and school work be suspended. But in the evening again, when there is usually a lull, a second period of instruction will be carried out. In well-equipped schools, to meet such conditions as these, it is customary to provide two complete and distinct staffs, both of instructors and mechanics. One staff takes the morning spell of work, while the second is held in readiness for the evening. This ensures that, both morning and evening, there shall be available for instruction a fresh, alert, and unfatigued staff.

A pupil will find that, as the first stage of his tuition, he is given the task of familiarising himself with the controls of a school biplane. The system we have described already, and a pupil should find no difficulty in mastering it. Placing himself in the driving-seat of the machine, while it is at rest on the ground, the pupil takes the upright lever in his right hand, and rests his feet on the rudder-bar, making the various movements of control, again and again, until he finds he is growing accustomed to them, and can place his levers in a position for an ascent or descent, or for a turn, without having to wait while he thinks what it is necessary to do.

In the next stage, a more interesting one, the pupil, occupying a seat immediately behind his instructor, is taken for a series of passenger flights. These accustom him to the sensation of being in the air, and also train his eye in judging heights and distances. A minor point the pupil should bear in mind, though his instructor will be quick to remind him, is not to wear any cap or scarf that may blow free in the rush of wind and become entangled with the propeller. Scarves need to be tightly wrapped; while it is usual, with a cap, to turn it with the peak to the back, and so prevent it from having a tendency to lift from the head. Many pupils provide themselves with a helmet designed to protect the head in case of an accident, and these are held firmly in position. Should a passenger's cap blow off, and come in contact with the propeller, it may be the cause of an accident. How carelessness may lead to trouble, in this regard, will be gathered from the following incident.

Some slight repairs had been made one day to the lower plane of a machine while it stood out on the aerodrome, and one of the workmen, through inadvertence, had left lying on the plane, near its centre, a roll of tape. The pilot decided to make another flight, and the motor was started and the machine rose. Suddenly the aviator was startled by a sound like a loud report, which seemed to come from the rear of his machine. The craft trembled for a moment, and he feared a structural collapse. Nothing worse happened, however, and he was able to pilot his machine in safety to the aerodrome. What had happened, it was then ascertained, was that the roll of tape, sucked back in the rush of wind, had been drawn into the revolving propeller and had broken a piece out of it. Luckily the impact had not been heavy enough to damage the propeller seriously, or cause it to fly to pieces.

A problem with which the pupil will be faced in his first flights, particularly if he is learning in winter, will be that of keeping himself warm. The speed at which an aeroplane travels, combined with the fact that it is at an elevation above the ground, renders the "bite" of the cold air all the more keen, and makes it difficult very frequently, even when one is warmly clad, to maintain a sufficient warmth in the body, and particularly in the hands and feet. The question of cold hands is, from a pilot's point of view, often a serious one. There is a case on record of an aviator who, his hands being so numbed that his fingers refused to move, found he could not switch off his motor when the time came to descend; and so he had to fly round above the aerodrome, several times, while he worked his numb fingers to and fro, and beat some life into them against his body. At last, having restored their circulation to some extent, he was able to operate the switch and make a landing. While on active service in winter, after flying several hours at high altitudes, and in bitter cold, the occupants of a machine have descended in such a numbed condition, despite their heavy garments, that it has been found necessary to lift them out of their seats. But a pupil need not face such hardships as these. He will be flying for short periods only, and at low altitudes; so if he makes a few wise purchases from among the selection of flying gear now available, and particularly if he equips himself with some good gloves, he should be able to keep sufficiently warm in the air, even if he is going through his training in winter.

A pupil will feel curious, naturally, as to his sensations in the first flights he makes with his instructor. Of the exact moment when the machine leaves ground he will be unaware probably, save for the cessation of any jolting or vibration, such as may be caused by the contact of the running wheels with the surface of the aerodrome. His first clearly-marked sensation, when in actual flight, will occur most likely when the pilot rises a little sharply, so as to gain altitude. Then the pupil will have a feeling one might liken to the ascent, in a motor-car, of a steep and suddenly-encountered hill; though in this case the hill is invisible, and there is no earth contact to be felt. This sensation of climbing is exhilarating; and when the pilot makes a reverse movement, descending towards the ground, the feeling is pleasant enough also, provided the dive is not too steep.

The pupil's chief sensation, probably, will be that of the rush of wind which beats against him. Some people feel this much more than others. There is sometimes a feeling--it is no more than temporary--of inconvenience and of shock. The pupil feels as though his breathing was being interfered with seriously; as though the pressure was so great he could not expel air from his lungs. But this sensation, even when it is experienced, is short-lived. In a second flight, quite often, the novice finds that this oppression diminishes very perceptibly; and soon he does not notice it at all. Motoring experience proves useful here, particularly high-speed driving on a track.

Some confusion is felt by the pupil, as a rule, and this is only natural, in regard to the pace at which the aeroplane travels through the air, and at the way in which the ground seems to be tearing away below. Occasionally, in a first flight, this impression of speed, and of height, produce in the pupil a sensation of physical discomfort; but it is one again which, in the majority of cases, is quickly overcome. A few balloon trips are a useful preliminary to flights in an aeroplane. They familiarise one in a pleasant way with the sensation of height, and accustom the eye also to the look of the ground, as it passes away below.

While he is making his first flights with the instructor, and apart from analysing his sensations, the pupil will observe the lever movements made by the pilot in controlling the machine; and the fact that will impress itself upon him, as he watches these movements, is that they are not made roughly or spasmodically, but are almost invariably gentle. During these flights as a passenger, and after he has accustomed himself to the novelty of being in the air, the pupil will be allowed by the instructor to lean forward and place his hand on the control lever; and in this way, by actually following and feeling for himself the control actions the pilot makes, he will gain an idea of just the extent to which the lever must be moved, to gain any specific result in the flight of the machine.

The next stage of tuition is that in which a pupil is allowed to handle a biplane alone, not in flight though but only in "rolling" practice on the ground--driving the machine to and fro across the aerodrome. The motor is adjusted so that, while it gives sufficient power to drive the machine on the ground and render the control surfaces effective, it will not permit the craft to rise into the air. This stage, a very necessary one, teaches the pupil, from his own unaided experience just what movements he must make with his levers to influence the control surface of the machine, and to maintain it, say, on a straight path while it runs across the ground. One of the discoveries he will make is that the biplane, if left to itself, shows a tendency to swerve a little to the left--the way the propeller is turning; but this inclination may be corrected, easily, by a movement of the rudder.

The pupil learns also to accustom himself, while in this stage, to the engine controls which have been explained already; and he is not likely to be guilty of the error of one excitable novice who, while driving his machine back on the ground towards the sheds at an aerodrome, after his first experience in "rolling" became so confused, as he saw the buildings looming before him, that he lost his head completely and forgot to switch off his motor. The result was that the aeroplane, unchecked in its course, crashed into some railings in front of the sheds and stood on its head. Not much damage was done however, and the novice was unhurt. He seemed as surprised as anyone at what had happened, and confessed that, for the moment, his mind had been an utter blank.

A pupil continues his practice in "rolling" till he can drive his machine to and fro across the aerodrome on a straight course, and with its tail raised off the ground; the latter action being obtained by the pupil by means of a suitable movement of the vertical lever which operates his elevating planes.

Now comes the time when a pupil, taking the pilot's seat, and with the instructor sitting behind him--so as to be ready, if necessary, to correct any error the novice may make--begins his first short flights across the aerodrome. He rises only a few feet to begin with, and flies on a straight course, alighting each time before he turns, and running his machine round on the ground. He repeats this test until his instructor feels he is sufficiently expert to take the machine into the air alone. When this stage is reached, the instructor leaves his position behind the pupil, and the latter goes on with his practice till he can fly the length of the aerodrome alone, landing neatly and bringing his machine round on the ground, and then flying back again to his starting point.

In the early days of flying schools, before a pupil went through any regular system of instruction, there were remarkable incidents in regard to these first flights. In one case a pupil, having bought his own aeroplane from the proprietors of a school, insisted on having installed in it a motor of exceptional power. When the time came for him to make his first flight alone, and he opened the throttle of this engine and it began to give its full power, the aeroplane ran only a short distance across the ground, and then leapt into the air. The engine was in charge of the machine, in fact, and not the pupil. Away above the aerodrome, and beyond its limits, in a strange, erratic flight, the biplane made its way. As the pupil struggled valiantly with his engine switch, which appeared to have become jammed, he made unconscious and jerky movements of his control levers. One moment the machine would ascend a little, the next it would approach nearer the ground; then it would swing either right or left. Those watching from the aerodrome held their breath. But with the luck of the beginner, a luck which is proverbial and sometimes amazing, the pupil managed at length to stop his motor and land without accident--though by no means gracefully--in an abrupt gliding descent.

Another story concerns one of those temperamentally reckless, happy-go-lucky men who, though providence seems to watch over them, are an anxiety nevertheless to their instructors. This pupil, breaking the rules of a school, flew out on one of his first flights beyond the limits of the aerodrome, disappearing indeed from the view of those near the sheds. Not far from the aerodrome lay a main road, with tramway-lines along it. A tram, with passengers on top, happened to be passing down the road; and it was to the astonishment of these passengers, and to their perturbation as well, that they observed an aeroplane in full flight, moving very low across a neighbouring field, and bearing down straight towards them. The machine passed, indeed, unpleasantly close above their heads, and then vanished as dramatically as it had appeared. Its pilot, as may be guessed, was the pupil who had disobeyed orders, and was now on a wild and erratic flight. Presently, after swerves and wanderings over the surrounding country, he was discerned making his way back towards the aerodrome, still flying unreasonably low. Some trees bordered one end of the aerodrome; and towards these, as though he meant to finish his exploit by charging into them, the novice was seen to be steering an undeviating course. Nearer he came to them, and still he did not turn either right or left. The instructor, and those gathered with him, made up their minds that nothing could avert an accident. But it happened that there was, between two of the trees, a space only large enough for an aeroplane to pass through. A skilled pilot, a man of experience, would not have cared to risk his machine in an endeavour to creep between those trees. But this pupil, a complete novice, steered boldly towards the opening and slipped through it with a precision that would have aroused the envy of an accomplished pilot. Then he landed on the aerodrome and climbed in leisurely fashion from his machine--"not having turned a hair," as the saying goes. The remarks of the instructor when he neared the machine, and began to unburden himself, do not appear to be on record, and no doubt this is as well.

Having shown his ability to make a succession of straight flights, taking his machine into the air with precision and landing without awkwardness, the pupil finds himself faced next with the problem of turning while in the air. On this stage, however, he is not allowed to embark alone. The instructor takes his place again in the passenger's seat, so as to be ready to help the novice should he become confused, or find himself in any difficulty. Turns to the left are attempted first; and the reason is that, the propeller of the aeroplane revolving to the left--and the motor too if it is a rotary one--the machine has a tendency which is natural to turn in this direction. Half turns only are tried at first, the pupil landing before he has completed the movement. In making these first turns a pupil finds that, apart from his action with the rudder-bar, it is necessary to employ the ailerons slightly, so as to prevent the biplane from tilting sideways. The outer plane-ends of the machine have indeed, when a turn is being made, a natural tendency to "bank" as it is called, or tilt upward; the reason being that, as the machine swings round, these outer plane-ends, moving faster for the moment than the wing-tips on the inside of the turn, exercise a greater lift, and have an inclination to rise. An experienced aviator, having learned what is a safe "banking" angle, makes a deliberate use of this tendency when he is turning, and may on occasion even exaggerate it, to facilitate the swing of his machine on a very rapid turn, and to prevent it skidding outwards. But with the novice, engrossed completely as he is with the mere problem of getting his machine round in the air, "banking" is an art that must be deferred for awhile. It is perilously easy, for a beginner, to overstep the danger-line between a safe "bank" and a side-slip.

It is not long before the pupil can make a full left-hand turn; and then he goes on to perfect himself in this movement, flying alone now, and repeating the turn till he feels he can make it with confidence, and at a fair height.

And now he comes to his final evolutions. Having mastered the left-hand turn, he proceeds to make one to the right. It used to be the contention--a contention that is now disputed--that in this movement, if the pupil employed his rudder-bar only, he would find the biplane showed an inclination to rise; a tendency due to the gyroscopic influence of the engine and propeller which--assuming a rotary engine is used--are now revolving in the opposite direction to that on which the machine is turned. What the pupil was recommended to do, in order to counteract this rising movement, was to tilt down his elevator a little, as he would in making a descent.

When right-hand turns can be made with the same facility as those to the left, the pupil begins to combine the two without descending, making left turns and right turns, and so achieving in the air a series of figures of eight. He learns also to fly a little higher, thus preparing himself for one of his certificate tests.

There are now certain very important rules which, in the navigation of his craft, he must accustom himself to bear constantly in mind. Should the engine of his machine, for example, betray any signs of failing, he must tilt down his elevator very promptly, and place his craft in a position for a descent. If he does not do this, and should the motor stop before he has his biplane at an angle for descent, the machine may lose speed so quickly, and its tail-planes show such a tendency to droop--owing to the lessening of pressure on their surfaces, consequent upon the failure of the motor--that there is a risk of the craft coming to a standstill in the air and then either falling tail-first, or beginning a side-slip that may bring it crashing to the ground.

The pupil must learn also, and this again is important, not to force his machine round on a turn while it is climbing. If he does so the power absorbed in the ascent, combined with the resistance of the turn, may so reduce the speed of the machine that it threatens to become "stalled," or reach a standstill in the air, with the result that it either side-slips or falls tail-first. The procedure the pupil is taught to follow is this: when he leaves the ground he climbs a little, then he allows his machine to move straight ahead; then he proceeds to ascend again for a spell, repeating afterwards the horizontal flight. In this way he ascends by a series of steps, like climbing a succession of hills in a car; and his turns should be made only during the spells when he is flying horizontally.

In this stage of his tuition, the pupil must learn also to make a vol-plan? or descent with his engine stopped. The essential point to be borne in mind, here, is that an aeroplane will continue in flight, and remain under control, even when it is no longer propelled by its engine. But what the aviator must do, should his engine stop through a breakdown, or should he himself switch it off, is to bring the force of gravity to his aid, and maintain the flying speed of his craft by directing it in a glide towards the ground. Provided he does this, and keeps his machine at such an inclination that it is moving at a sufficient speed through the air, he will find that the craft maintains its stability and that he has full command over its control surfaces, being able to turn, say, right or left, or either increase or slightly decrease the steepness of his descent. But all the time, of course, seeing that it is gravity alone which is giving him his flying speed, he is obliged to plane downward.

A vital point to remember, when a pupil is handling a "pusher" type of biplane, is to incline the machine well downward, by a use of the elevator, before switching off the motor. Unless this is done, and if the machine is, say, at its normal horizontal angle when the engine is stopped, the sudden removal of pressure from the tail-planes of the craft, brought about by the absence of the wind-draught from the propeller, may cause the tail so to droop as to render inoperative any subsequent action of the elevator. When the tail droops, the main-planes are set at a steep angle to the air, and this has a slowing-up influence on the whole machine. It threatens therefore to stand still in the air; its controls become useless; and the pupil is faced probably with the danger of a side-slip.

A story will illustrate this point; and it is one that has a special significance, seeing that the error which might have cost him his life was made by an aviator of experience. He had learned to fly on a monoplane, and had devoted his subsequent flying, for many months, to this one type of machine. Then he found himself associated with an enterprise in which a number of "pusher" biplanes were employed, and he decided that it would be useful for him to become accustomed to this type of machine. His flying experience of course helped him, and he soon found himself passing to and fro above the aerodrome, the biplane well in hand. Then he thought he would make a vol-plan? with his motor stopped, as he had been in the habit of doing in a monoplane. He switched off his engine without further thought, and moved his elevator to a position for the descent. But it was here that he made the mistake. In a monoplane, which has the weight of the engine and other gear well forward in the machine, the bow has a natural tendency to tilt down when the motor is cut off--particularly as the propeller-draught ceases to sweep under the sustaining planes. Therefore one can, in such a machine, switch off safely without first shifting the elevator, and getting the bow down as a preliminary. What the pilot had forgotten, for the moment, was the essential difference between monoplane and biplane. When he had switched off the engine in the biplane, and moved his elevator as he was accustomed to do, he found to his dismay that the machine failed to respond. Instead of pointing its bow down, indeed, it began to tilt rearward. Also, and this fact was noted by the airman with even more dismay, the craft lost forward speed so rapidly that it became uncontrollable. The next moment, the pilot helpless in his seat, the machine began a side-slip towards the ground. One sweep it made sideways, falling till it was not far short of the surface of the aerodrome. It paused an instant, then began a side-slip in the opposite direction. But here good fortune came to the pilot's aid. In this second swing, the machine being near the ground, it came in contact with the surface of the aerodrome before the "slip" had time to develop any high rate of speed. The biplane took the ground sideways, breaking its landing-chassis and damaging the plane-ends which came first in contact with the earth. But the pilot emerged from the wreckage unhurt. The accident was a lesson to him, though, as it was to others, and as it should be to all pupils. A machine must be in a gliding position before the engine is switched off.

The art of the accomplished pilot, granted there is no reason for him to reach earth quickly, is to glide at as fine an angle as is possible, consistent of course with maintaining the speed of the machine through the air, and so preserving his command over its controls. A beautifully-timed, fine glide, the machine stealing down gracefully, and touching the aerodrome light as a feather, at a precise spot the airman has decided on even when he was several thousand feet high, is a delightful spectacle for the onlooker, and a keen pleasure also--from the point of view of his manipulative skill--to the aviator himself. But a pupil, at any rate in his first attempts, must not concern himself too much with any idea of a fine or graceful glide. It is his business to get to the ground safely, and not trouble too much whether his method is accomplished, or merely effective. Once with the bow of his machine down, and his motor switched off, it must be his concern to maintain the forward speed of his machine, which can be done only by holding it well on its dive. For the novice, if he attempts any fine or fancy gliding, there is the very real danger that, in his inexperience, he may lose forward speed to such an extent that his controls become inoperative, and his machine threatens to side-slip. One's ear should, apart from the inclination of the machine, and the sensation of the descent, help one materially in judging the speed of a glide. There is a "swish" that comes to the ear, now the engine is no longer making its clamour, which gives a guide to the pace of one's downward movement. Aviators who are skilled, and have done a large amount of flying, are able to judge with accuracy, by the ear alone and without the aid of a mechanical indicator, what their speed is as they pass through the air.

Having held his machine firmly on its glide, till it is quite near the surface of the aerodrome, the pupil has next to think of making a neat contact with the ground. The art here is, at a moment which must be gauged accurately, to check the descent of the machine by a movement of the elevator--to "flatten out," as the expression goes. If the movement is made neatly the craft should, when only a few feet from the ground, change from a descent into horizontal flight, and continue on this horizontal flight for a short distance, losing speed naturally each moment--seeing that there is no driving power behind it--and so losing altitude also through its decrease in speed, until its wheels come lightly in contact with the ground, and it runs forward and then stands still. What the novice may do, if he is not careful, is to "flatten out" when he is too high above the ground. The result is that the machine slows up till it stands still in the air, robbed of its speed, and then makes what is called a "pancake" landing: it descends vertically, that is to say, instead of making contact with the ground at a fine angle and with its planes still supporting it; and the effect of such a "pancake," if the machine comes down with any force, may be that the landing-chassis is damaged, or perhaps wrecked. But as a rule, remembering that he has careful instruction to guide him before he attempts a gliding descent, the pupil masters the art of landing without difficulty, and without mishap.

Now, after repeating perhaps certain of his evolutions, at the discretion of his instructor, in order to make sure that he can accomplish them with ease, the pupil is ready for the tests which will give him his certificate of proficiency.

CHAPTER VI

THE TEST FLIGHTS

The sport of aviation is controlled throughout the world, and flying tests and events of a competitive character are governed, by the International Aeronautical Federation. To the deliberations of this central authority are sent delegates from the Aero Clubs of various countries; and to these Aero Clubs, each in its respective country, falls the task of governing flight, according to the rules and decisions of the central authority. In Britain, controlling aviation in the same way that the Jockey Club controls the Turf, we have the Royal Aero Club of the United Kingdom; and it is this body, acting in its official capacity, which grants to each new aviator, after he has passed certain prescribed tests, a certificate which proclaims him a pilot of proved capacity, and without which it is impossible for him to take part in any contests held under the auspices of the Club. The certificate, which is of a convenient size for carrying in the pocket, contains a photograph of the pilot for purposes of identification, and specifies also the rules under which the certificate is issued and held.

The theory of these tests, as imposed by the Club before it grants its certificates, is that the novice should--so far as is possible in one or two flights, made over a restricted area, and in a limited space of time--be called on to show that he has a full control over a machine in what may be called the normal conditions of flight. He is asked to ascend, for instance, and gain a fair flying altitude; then to make such evolutions as will demonstrate his command over the control surfaces of the machine; and finally to show that he can, with his motor switched off, descend accurately in a vol-plan? and bring his machine to a halt within a specified distance of a mark. The tests are set forth, officially, as follows:--

A and B. Two distance flights, consisting of at least 5 kilometres (3 miles 185 yards) each in a closed circuit, without touching the ground; the distance to be measured as described below.

C. One altitude flight, during which a height of at least 100 metres (328 feet) above the point of departure must be attained; the descent to be made from that height with the motor cut off. The landing must be made in view of the observers, without re-starting the motor.

The rules drafted by the Club to govern these flights are set forth herewith:--

The candidate must be alone in the aircraft during the tests.

The course on which the aviator accomplishes tests A and B must be marked out by two posts situated not more than 500 metres (547 yards) apart.

The turns round the posts must be made alternately to the right and to the left, so that the flights will consist of an uninterrupted series of figures of eight.

The distance flown will be reckoned as if in a straight line between the two posts.

The alighting after the two distance flights in tests A and B shall be made:--

(a) By stopping the motor at or before the moment of touching the ground.

(b) By bringing the aircraft to rest not more than 50 metres (164 feet) from a point indicated previously to the candidate.

All alightings must be made in a normal manner, and the observers must report any irregularity.

These flights as specified to-day, though they present no difficulty to the pupil who has been well trained, are more stringent than they were in the first scheme of tests as prescribed by the Club, and as enforced for several years. In those early rules the distances were the same as they are to-day, but in the altitude flight the height required was only 50 metres (164 feet)--just half the height specified to-day. It was not laid down, either, in the first rules, that the engine should be stopped in this altitude flight when at the maximum height, and that the descent should be made in a complete vol-plan? without once re-starting the motor. As originally framed, indeed, the rule as to the control of the engine in this altitude test was the same as in regard to the distance flights--i.e., that it should be stopped "at or before the moment of touching the ground." What the present rule means, in this respect, is that the pupil must be really proficient at making a vol-plan? without any aid at all from his engine, before he can hope to pass the test;

and such a proved skill--say in the making of his first cross-country flight, should his engine fail suddenly--may spell the difference between a safe or a dangerous landing.

The test flights for the certificate, undertaken only in such weather conditions as the pupil's instructor may think suitable, are watched by official observers appointed by the Royal Aero Club. It is the business of these observers, when the prescribed flights have been made, to send in a written report concerning them to the Club; and acting on this report, after it has been considered and shown to be in order, the Club issues to the pupil his numbered certificate. With the successful passing of his tests the pupil's tuition is at an end. He is regarded no longer as a novice, but as a qualified pilot.

CHAPTER VII

PERILS OF THE AIR

There are people, very many people, who still regard flying as an undertaking of an unreasonable peril, essayed mainly by those who are in quest of money, notoriety, or sensation at any price. Such people--still to be met with--have one mental picture, and one only, of the flight of an aeroplane. They imagine a man in the air--and this mere idea of altitude makes them shudder; and they picture this man in a frail apparatus of wood and wire, capable of breaking to pieces at any moment; or even if it does not break, needing an incessant movement of levers to maintain it in a safe equilibrium; while they reckon also that, should the engine of the machine suffer any breakdown, the craft will drop to earth like a stone. Prejudice dies hard; harder no doubt in England than in other countries. There are still people, not few of them but many, who would be ready to declare, offhand, that one aeroplane flight in six ends in a disaster.

It is a truism, but one that has a peculiar truth in aviation, to say that history repeats itself. To-day we find large numbers of people who still cherish the opinion that--save perhaps when on service in war--it is nothing less than criminal foolishness for men to ascend in aeroplanes. That attitude of mind persists; the growing safety of flight has not affected it to any appreciable degree. But those eager for the progress of aviation need not despair, or imagine that their particular industry is being treated with any exceptional disapprobation. They have only to look back a little in our history, no great distance, and read of the receptions that were accorded the first pioneers of our railways. Public meetings of protest have not been held to condemn aviation; yet they were frequent in the days when the first railways were projected. Vast indignation was indeed aroused; it was declared to be against all reason, and a matter of appalling risk, that people should be asked to travel from place to place in such "engines of destruction." But the railways managed to survive this storm. They were placed here and there about the country; they were improved rapidly; and it would be hard, to-day, to find a safer place than the compartment of a railway train.

Motor-cars, when their turn came, had to go through a similar ordeal. There was the same indignation, the same chorus of protest; and when the first of the pioneers, greatly daring, began actually to drive their cars on the public highway, there were people who believed, and who declared forcibly, that to permit such machines on our roads was the crime of the century. Had not these pioneers struggled valiantly, sparing neither time nor money, it is possible that the motor-car might have been driven from the highway. But here again progress, though it was retarded, could not be checked. The motor-car triumphed. It grew rapidly more reliable, more silent, more pleasing to the eye; and to-day it glides in thousands along our roads, a pleasure to those who occupy it, a nuisance neither to pedestrians nor to other wheeled traffic; more under control when it is well driven, and more ready to stop quickly when required, than any horsed vehicle which it may have replaced. At one time the papers were full of such headlines as: "Another Motor-car Accident." Each small mishap received prominent attention: and to the majority of people it seemed the wildest folly to travel in such vehicles. Yet to-day--such is progress--these same people ride in a motor-car, or a motor-cab, quite as a matter of course and without a thought of risk.

When one discusses flying and its dangers, it is essential to maintain an accurate sense of proportion. In the very earliest days, for instance, it must be realised that the few men who then flew--they could be numbered on the fingers of one hand--exercised the greatest caution. They did not fly in high winds; they treated the air, realising its unknown perils, with a very great and a very commendable respect. Thus it was that thousands of miles were flown, even with the crudest of these early machines, and with motors that were constantly giving trouble, without serious accident. But after this, and very quickly, the number of airmen grew. New aviators appeared every day; contests were organised extensively; there were large sums of money to be won, provided that one pilot could excel another. And the spirit of caution was abandoned. Even while they were still using purely experimental machines--craft of which neither the stability nor the structural strength had been tested adequately--there grew a tendency among airmen to fly in higher winds, to subject their machines to greater strains, and to attempt dangerous manoeuvres so as to please the crowds who paid to see them fly.

It was not surprising, therefore, that flying entered upon an era of accidents. Such disasters were inevitable--inevitable, that is to say, in view of the tendencies that then prevailed; though it is a melancholy reflection that, had men been content to go ahead with the same slow sureness of the pioneers, many of those lives which were lost could have been saved.

To the public, not aware exactly of all that was going on, it appeared as though the navigation of the air, instead of growing safer, was becoming more dangerous. There were suggestions, indeed, made quite seriously and in good faith, that these endeavours to fly should cease; that the law should step in, and prevent any more men from risking their lives. What people failed to realise, when they adopted this view, was that instead of one or two men flying there were now hundreds who navigated the air; that flights in large numbers were being made daily; that thousands of miles instead of hundreds were being traversed by air--and often under conditions the pioneers would have considered far too dangerous. These facts, had they been realised, would have shown people what was actually the true state of affairs; that, though accidents seemed numerous, and were indeed more frequent than they had been in the earliest days of flying, they were as a matter of proportion, reckoning the greater number of men who were flying, and the thousands of miles which were flown, growing steadily less frequent.

There was this important fact to be reckoned with also. Each accident that happened taught its lesson, and so made for future safety. A considerable number of those early accidents can, for instance, be traced to some structural weakness in a machine. The need in an aircraft then, as now, was lightness; and in those days designers and builders, owing purely to their inexperience, had not learned the art, as they have to-day, of combining lightness with strength. So it was that, as more powerful motors began to be fitted to aeroplanes, and greater speeds were attained, it happened sometimes, when a machine was being driven fast through a wind, that a plane would collapse, and send the machine crashing to the ground; or in making a dive, perhaps, either of necessity or to show his skill, a pilot would subject his machine to such a strain that some part of it would break.

From such disasters as a rule, greatly to be regretted though they were, the industry emerged so much the wiser. The strength of machines was increased; the engines which drove them were rendered more reliable; and gradually too, though none too rapidly, the airmen who piloted them grew in knowledge and skill. But all this time, while flying was being made more safe, there were accidents frequently for the papers to report; and this was due entirely to the fact that there were now thousands of men flying, where previously there had been fifties and hundreds. The public could not realise how rapidly the number of airmen had grown; that practically every day, at aerodromes scattered over Europe, flights were so frequent that they were becoming a commonplace. It was in 1912, as one of its many services to aviation that the Aero Club of France was able to show, by means of statistics which could not be questioned, that for every fatality which had occurred in France, during that particular year, a distance of nearly 100,000 miles had been flown in safety.

The cause of many of the early accidents was, as we have suggested, the breakage of some part of a machine while in flight. In an analysis for instance of thirty-two such disasters, it was shown that fourteen were due to the collapse of sustaining planes, control-surfaces, or some other vital part of a machine. And this risk of breakage in the air was increased, in many cases, by the building of experimental machines by men who had no qualifications for their task, and who erred only too frequently, in their desire to attain lightness, on the side of a lack, rather than an excess, of structural strength.

There are many cases, unfortunately, that might be cited; but one may be sufficient here. A man with an idea for a light type of biplane, a machine designed mainly for speed, had an experimental craft built--this was in the pioneer days of 1909--and insisted on fitting to it a motor of considerable power. It was pointed out to him that his construction was not sufficiently strong, in view of the speed at which his machine would pass through the air. But he was of the quiet, determined, self-opinionated type, who pursued his own way and said little. He did not strengthen his constructional, and he began a series of flying tests. In the first of these, which were short, the planes stood up to their work, and the fears of the critics seemed groundless. But a day came when, venturing to some height, the aviator encountered a strong and gusty wind; whereupon one of his main-planes broke, and he fell to his death.

As a contrast to this tragedy, and a welcome one, there is a humorous story, that is true, told of one experimenter. His knowledge of construction was small, but what he lacked in this respect he made up for in confidence; and he built a monoplane. This was in the days just after the cross-Channel flight, and experimenters all over the world were building monoplanes, some of them machines of the weirdest description. The craft built by this enthusiast seemed all right in its appearance; nothing had been spared, for instance, in the way of varnish. When wheeled into the sun, for its first rolling test under power, it looked an imposing piece of work. Friends were in attendance, photographers also; and the would-be aviator was in faultless flying gear. Mounting a ladder, which had been placed beside the machine, he allowed his weight to bear upon the fusilage, and proceeded to settle himself in his seat. But he, and the onlookers, were startled as he did so by an ominous cracking of wood. It grew louder; something serious and very unexpected was happening to the machine. As a matter of fact, and just as it stood there without having moved a yard, the whole of the flimsy structure parted in the middle, and the machine settled down ignominiously upon the ground, its back broken, and with the discomfited inventor struggling in the debris.

It was far from easy, in the early days, for even an expert constructor to calculate the strains encountered under various conditions of flight. In wind pressure, under certain states of the air, there are dangerous fluctuations-- fluctuations which, even with the knowledge we possess to-day, and this is far from meagre, exhibit phenomena concerning which much more information is required. Machines have collapsed suddenly, while flying on a day when the wind has been uncertain, and have done so in a way which has suggested that they had encountered, suddenly, a gust of an altogether abnormal strength. Occasionally, though research work in this field is extremely difficult, it has been possible to gain data as to the existence of conditions, prevalent as a rule over a small area, which would spell grave risk for any aeroplane which encountered them. There is a strange case, verified beyond question, which occurred during some tests with man-lifting kites at Farnborough. These kites are strongly built, and withstand as a rule extremely high winds. On this particular day a kite, when it had reached a certain altitude, was seen to crumple up suddenly. The wind did not seem specially strong--not at any rate on the ground; and there appeared no reason for the breakage of the kite. Another was sent up; but the same thing happened, and at the same altitude. Then the officer who was in charge of the kites sent for a superior. A third kite was flown to see what would happen. This one broke exactly as the others had done, and at just the same height--about five hundred feet. Precise data could not be gained as to this phenomenon; but the breaking of these kites--which had withstood extremely high pressure in previous tests--was reckoned to be due to the fact that, when they reached a certain point in the air, they were subjected to the violent strain of a sudden and complete change in the direction of the wind. To the pilot of an aeroplane, entering without warning some such area of danger, the result might naturally be serious in the extreme.

The air has been, and is still, an uncharted sea. It does not flow with uniformity over the surface of the earth. It is a constantly disturbed element, and one that has the disadvantage of being invisible. An aviator cannot see the dangerous currents and eddies into which he may be steering his craft; and so it was not surprising, in those days when aircraft were frailer than they should have been, and cross-country flights were first being made, that machines broke often while in flight and that the airman's enemy, the wind, claimed many victims.

Wind fluctuations that are dangerous, those which possess for one reason or another an abnormal strength, are encountered frequently when a pilot is fairly near the earth; and his peril is all the greater in consequence. On a windy day, one on which there are heavy gusts followed by comparative lulls, it is when he is close to the ground, either in ascending or before alighting, that a pilot has most to fear. If he is well aloft, with plenty of air space beneath him, and particularly if he has a machine that is inherently stable, he has little to fear from the wind; save, perhaps, should his engine fail him, or should he find--as has been the case in war flying--that the force of the wind, blowing heavily against him, and reducing the speed of his machine, has prevented him from regaining his own lines before his petrol has become exhausted. The modern aeroplane, when its engine-power is ample, and it is at a suitable altitude, can wage battle successfully even with a gale. But it must rise from the earth when it begins a flight, and return to earth again when its journey is done; and here, in the areas of wind that are disturbed by hills, woods, and contours of the land, there are often grave dangers. The wind at these low altitudes blows flukily. Its direction may be affected, for instance, owing to the influence of a hill or ridge. A side gust, blowing powerfully and unexpectedly against a machine, just as it is nearing the ground before alighting, may cause it to tilt to such an angle that it begins a side-slip. If the craft was sufficiently high in the air, when this happened, the pilot would be able, probably, to convert the side-slip into a dive, and the dive into a renewal of his normal flight. But if such a side-slip begins near the ground, and there is an insufficient amount of clear space below the machine, it may strike the ground in its fall, and become a wreck, before there is time for the pilot, or for the machine itself, to exercise a righting influence. The fact that a craft may be forced temporarily from its equilibrium, say by a side-slip, is known now to represent no great risk for the airman, granted always that he has the advantage of altitude. The machine, in such circumstances, falls a certain distance. This is inevitable, and for the reason that it must regain forward speed--which it has lost temporarily in its side-slip--before its own inherent stability can become effective, or its pilot regain influence over his controls. And it is this unavoidable descent, this short period during which the machine is recovering its momentum, and during which the pilot has no power of control, that represents in a heavy wind the moments of peril, should a pilot enter an area of disturbance just as he nears the ground.

An aeroplane, when it sets out to fly in bad weather, may be likened to a boat that is being launched from a beach upon a rough and stormy sea. It is the waves close inshore, which may raise his craft only to dash it to destruction, that the boatman has chiefly to fear; and for the aviator, when he leaves the land and embarks upon the aerial sea, or when he returns again from this element and must make his contact with the earth, there lurks a risk that, caught suddenly by an air wave, and with insufficient space beneath his machine, he may be forced into a damaging impact with the ground. But the skill of designers and constructors, to say nothing of the growing experience of aviators, is working constantly towards a greater safety.

Of the risk attached to engine failure, when he is piloting a craft fitted with only one motor, an airman is reminded frequently, not only from his own experience, but from that of other flyers. With the aeroplane engine, even with types that have gained a high average of reliability, there are many possibilities of a slight mishap--each of them sufficient, for the moment, to put an engine out of action--that the pilot who is flying across country must, all the time he is in the air, have at the back of his mind the thought that at any moment, and perhaps without any warning, he may find that his motive power has gone. A magneto may fail temporarily; an ignition wire or a valve spring break. The aeroplane engine of to-day is, of course, an infinitely more reliable piece of apparatus than it was in those early days when Henry Farman, working with extraordinary patience at Issy-les-Moulineaux, was endeavouring--and for a long time without success--to make the motor in his Voisin biplane run for five consecutive minutes without breakdown. The war has shown us, and under working conditions which have been exceptionally trying, how reliable the aero-motor has become. But until duplicate plants have been perfected, and more than one motor is fitted to aircraft as a matter of course, there must always be this risk of failure.

In the mere stoppage of a motor no great danger is implied. The pilot must descend; that is all. His power gone, he must glide earthward. But where the risk does lie, in engine failure, is that it may occur at a moment when the airman is in such a position, either above dangerous country or while over the sea, that he cannot during his glide reach a place of safety. A study of flying will show how awkward, and how perilous on many occasions, has been the stoppage of a motor while a machine is in the air. Two historic instances, though they did not, fortunately, end in a loss of the pilot's life, were the compulsory descents into the Channel made by the late Mr. Hubert Latham, during his attempts, in 1909, to fly from Calais to Dover. In both these cases-- once when only a few miles from the French shore, and on the second occasion when the aeroplane was quite near its destination--the motor of the Antoinette monoplane failed suddenly, and the aviator could do nothing but plane down into the water. On the first occasion he alighted neatly, suffering no injury, and being rescued by a torpedo boat; but in the second descent, striking the water hard, he was thrown forward in his seat and his head injured by a strut.

Less fortunate, in a case of presumed engine failure that will become historic, was Mr. Gustave Hamel. Eager to reach Hendon, so as to take part in the Aerial Derby on May 23rd, 1914, his great experience of Channel flying induced him to risk the crossing with a motor which, on his flight from Paris to the coast, had not been running well. His monoplane was a fast machine, and the flight across Channel would have taken him less than half an hour. But at some point during the crossing, it seems obvious, his engine failed him, and he was unable to prolong his glide either to gain the shore, or the vicinity of a passing ship. His monoplane was never recovered; but the body of the aviator--whose loss was mourned throughout the flying world and by the general public as well--was discovered by some fishermen while cruising off the French coast, and identified by means of a map, clothing, and an inflated motor-cycle tyre; the last-named being carried by the airman round his body to act as an improvised life-belt.

Engine failure, though a fruitful cause of minor accidents, and of the breakage of machines, has led to few fatalities; and this has been due very largely to the fact that, though machines have descended under dangerous circumstances, and have been wrecked in a manner that would appear almost certain to kill their occupants, the pilots and passengers have, as a matter of fact, escaped often with no more than a shock or bruises. An aeroplane does not strike the ground with the impact of a hard, unyielding structure. It is essentially frail in its construction; and this frailness, though it spells destruction for the machine in a bad descent, provides at the same time an element of safety for its crew. Take the case for instance of a machine falling sideways, and striking the ground with one plane or planes. These planes, built of nothing stronger as a rule than wood, crumple under the impact. But in their collapse, which is telescopic and to a certain extent gradual, a large part of the shock is absorbed. By the time the fusilage which contains the pilot touches ground, the full force of the impact is gone. And it is the same, often, if a machine makes a bad landing, say on awkward ground, and strikes heavily bow-first. Granted that the occupants of the machine are well-placed, and prevented by retaining belts from being flung from the machine, they should escape injury from the fact that there is so much to be broken, in the way of landing-gear and other parts, before the shock of the impact can reach them in their seats.

Had it not been for the capacity of the aeroplane to alight in awkward places without injury to its pilot, many lives might have been lost through descents in which motors have failed. Aviators have been obliged to land in most unsuitable places: on the roofs of houses, for instance, in small gardens, and frequently on the tops of trees. If he finds his engine fail him when he is over a wood or forest, and there is no chance save to descend upon the trees, a skilled pilot may save himself as a rule from injury. Planing down, till he is just above the tree-tops, he will then check suddenly, by a movement of his elevator, the forward speed of his machine. The craft will come to a standstill in the air; then, the support gone from its planes owing to the loss of forward speed, it will sink down almost vertically, and with very little violence, on to the tops of the trees. The machine itself will naturally be damaged, seeing that boughs will pierce its wings in many places, and that one or more of its planes may possibly collapse. But the net result of such a landing--and this is the point which is important for the pilot--is that the machine will be caught up and suspended on the trees, making a comparatively light and gradual contact, instead of there being any risk of its driving through the trees and making a heavy impact with the ground.

Humour, sometimes, may be extracted from such a predicament as engine failure, though it needs an aviator with a very deeply ingrained sense of humour to do so. The story is told, however, of a pilot who, flying across difficult country with a passenger, found that his motor failed--as they often will--just at a moment when there seemed no possible landing-point below. Looking over the side of his machine, and glancing quickly here and there, the aviator saw no alternative but to bring his craft down in an orchard that lay below. Pointing downward, to acquaint his passenger with their unpleasant situation, and to call his attention also to the orchard, the pilot said with a smile:

"I hope you're fond of apples!"

There is a risk in engine failure which has been emphasised more than once; and it is that which may attend the pilot who, while prolonging a glide in order to reach some landing-point, may be struck by a gust, or enter some area of disturbed wind, just before he reaches the ground and while his machine, moving slowly, is not in a position to respond effectually to its controls. In one case an aviator, struggling back towards the aerodrome with a motor which was not giving its power, found that it stopped suddenly when he was not far from a wood. Beyond the wood, which stood on a ridge, there was a stretch of grassland. Endeavouring to reach this promised landing-point, and holding his machine on a long glide, the airman came across above the trees. He had almost reached his goal when his machine entered a sudden down-current of wind--occasioned, no doubt, by the proximity of the trees and ridge. Caught by this eddy, with no motive power to help him and very little speed on his machine, the pilot could not check its sudden dive; and the craft struck ground so heavily that both he and his passenger were killed.

We have mentioned previously, as a fruitful cause of accident, that structural weakness of machines which has led, when conditions have been unfavourable, to a sudden collapse in the air. But apart from weakness in construction, and notably in accidents with early-type machines, there was the risk attached to mistakes in design, which produced machines which were unstable under certain conditions--and the dangers also which were due to inefficient controlling surfaces. It was no uncommon thing, in pioneer days, for a machine to be built which would not respond adequately to its elevator or rudder; though this unpleasant fact might not be discovered by the pilot until he was actually in flight, and perhaps at some distance from the earth. In one case, which is authenticated, a two-seated monoplane of a new type was tested at first in a series of straight flights, and found to be promising in its behaviour. A skilled pilot then took charge of it, and, carrying a passenger, proceeded to some more ambitious flights. Steering the machine away across the aerodrome, and flying at a low elevation, he approached a belt of woods. The machine was too near the ground to pass over the tops of the trees; so the aviator decided to make a turn, and fly parallel with the wood. But when he put his rudder over, so as to bring the machine round in a half-circle, he found to his dismay that there was no response. In the design of the machine, as it was found afterwards, the rudder had been made too small: it would not steer the machine at all. In the little space that was left him, and to avoid crashing into the trees, the pilot had to bring his craft to earth in such an abrupt dive that it was wrecked completely. He and the passenger, though, escaped unhurt.

Carelessness has, fairly frequently, played its part in aeroplane disasters. Sometimes a pilot has been careless, or perhaps in a hurry, and has failed to locate some defect which, had it been seen and attended to, would have saved a disaster when a machine was in flight. Such inattention, which is sufficiently dangerous in the handling of any piece of mechanism, is deadly in its peril when those who are guilty of it navigate the air. A man who brings out a machine time after time, and ascends without examining it carefully, is adding vastly to the risks that may attend his flight; and the same remark will apply to the carelessness of mechanics; though as a class, in view of the arduous nature of their work, and of the long hours they have frequently to be on duty, with no more than hasty intervals for rest, their average of care and accuracy is very high. But there have been cases--mostly in the past though--in which a machine has developed a structural defect, or some defect say in its control gear, which ought to have been observed by its mechanics, but which has not been so detected, and has led to a catastrophe in flight. With machines built lightly, and subjected to heavy strains when at high speeds, it is vital that the inspection of such craft, that the examination of every detail of them, should be carried out in a spirit of the greatest care. The fraying through of a control wire, unnoticed by those in charge of a machine, has been sufficient to cause a disaster; while carelessness in overhauling a motor, a task of supreme importance, seeing that its engine is the heart of an aeroplane, has been another cause of accident. It is vital that, when an airman ascends, both his machine and his motor should be in perfect working trim. He himself, before he flies, and after his aeroplane has been wheeled from its shed, should make it a habit to look over the machine, so as to impose his own personal check upon the work his mechanics have done.

Even when every care has been taken, and a machine ascends in perfect trim, there is the human factor, represented by the pilot, which must be considered always in a study of aeroplane accidents. There is often, when a catastrophe seems imminent, a choice of things that may be done. If an engine fails, for instance, under awkward circumstances, the pilot may have, say, three courses open to him in regard to his descent. Two may spell disaster and the third safety. It is here that the innate judgment of a pilot, combined with his experience, will tell its tale. But this personal element in flying, and particularly in regard to an accident, is often a very difficult one for which to make allowances.

The whole problem of aeroplane disasters is, to the analyst, one of unusual complexity. Take for example the case of a pilot who is flying alone in his machine, and at an altitude of several thousand feet. Suddenly something happens; the machine is seen to fall and the pilot is killed. Experts come to examine the aircraft, but it is wrecked so completely that little which is reliable can be gathered from any inspection; while the man who could explain what has happened--the pilot of the machine--is dead. The statements of eyewitnesses, when taken on such occasions, are often misleading. One person heard a crash, and saw something fall away from the machine. Another declares the engine stopped suddenly and that the machine "fell like a stone." Another says he is sure he saw one of the wings fold upwards and the machine swing and fall. And so on. It is extremely difficult, even for a technical eye-witness, to be sure of what he sees when things happen quickly and at a distance from him; while the statements of non-technical people, who are not trained in observation, are generally so unreliable as to be useless.

It has happened often therefore, far too often, in aeroplane fatalities that have happened from time to time, that the cause of such accidents has, even after the most careful investigation, had to be written down a mystery. But in more than a few cases, though the evidence has been far from conclusive, it has been considered that a pilot has been guilty of some error of judgment. There were puzzling instances, notably in the early days of flying, when airmen began first to make cross-country flights, of engines being heard to fail suddenly, and machines seen to fall to destruction. That engines should break down was not surprising; they were doing so constantly; but there was no reason why, even if they did fail, a machine should fall helplessly instead of gliding. But what was thought to have happened, in more than one of these cases, was that the pilot, through an error of judgment, had failed to get down the bow of his machine when his motor gave signs of stopping. The craft concerned were, it should be mentioned, "pusher" biplanes; and the same rule applied to them, in cases of engine failure, as has been explained in a previous chapter, and as is emphasised nowadays in the instruction of the novice. But in those days the beginner had frequently to learn, not from wise tuition, but from bitter experience; and he was lucky, often, if he learned his lesson and still retained his life. On certain early-type biplanes, for instance, machines with large tail-planes, and engined as a rule by a motor which was giving less than its proper amount of power, it was most dangerous for a pilot if, on observing any signs of failing in his engine, he sought to fly on in the hope that the motor would "pick up" again, and continue its work. Directly there was a tendency of the motor to miss-fire, or lessen in the number of its revolutions per minute, the consequent reduction of the propeller draught, as it acted on the tail of the machine, would cause this tail to droop, and the machine to assume very quickly a dangerous position. And when once it began to get tail-down, as pilots found to their cost, there was nothing to be done. The machine lost what little forward speed it had, and either fell tail-first, or slipped down sideways. Such risks as these, which were very real, were rendered worse owing to the fact that, in much of the cross country flying of the early days, pilots flew too low. They lacked the confidence of those who followed them, and were too prone to hug the earth, instead of attaining altitude. It was not realised clearly then, as it is now, that in height lies safety. And so when a machine lost headway through engine failure, and was not put quickly enough into a glide, it happened often that it had come in contact with the earth, and had been wrecked, before there was any chance for the pilot to regain control, or for the machine itself to exhaust its side-slip, and come back to anything like a normal position.

But the failure of the human factor in flying, the lack of skill of a pilot that may lead to disaster, is shown by statistics to play no more than a small part, when accidents are studied in numbers and in detail. Some time before the war, in an analysis of the accidents that had befallen aviators in France-- accidents concerning which there was adequate data--it was shown that only 15 per cent. of them could be attributed to a failure in judgment or skill on the part of the pilot.

Apart from errors, however, in what may be called legitimate piloting, there have been regrettable accidents due to trick or fancy flying. Putting a machine through a series of evolutions, to interest and amuse spectators, is not of course in itself to be condemned. In such flying, and notably for instance in "looping the loop," facts were learnt concerning the navigation of the air, and as to the apparently hopeless positions from which an aeroplane would extricate itself, which were of very high value, from both a scientific and practical standpoint. Public interest in aviation was increased also by such displays; and it is very necessary that there should be public interest in flying, seeing that it is the public which is asked to pay for the development of our air-fleets. But the man who undertakes exhibition flying needs not only to be a highly-skilled pilot, but a man also of an exceptional temperament--a man whose familiarity with the air never leads him into a contempt for its hidden dangers; a man who will not, even though he is called on to repeat a feat time after time, abate in any way the precautions which may be necessary for his safety. In looping the loop, for instance, or in upside-down flying, it is necessary always that the aeroplane should be at a certain minimum height above the ground. Then, should anything unexpected happen, and the pilot lose command temporarily over his machine, he knows he has a certain distance which he may fall, before striking the ground; and during this fall the natural stability of his machine, aided by his own operation of the guiding surfaces, may bring it back again within control. But if he has been tempted to fly too near the ground, and has ignored for the moment this vital precaution, and if something happens for which he is not prepared, then the impact may come before he can do anything to save himself.

In the early days of flying, when aviators attempted an acrobatic feat, they ran a far heavier risk than would be the case to-day; and for the simple reason that their machines, not having a strength sufficient to withstand any abnormal stresses, were likely to collapse in the air if they were made to dive too rapidly, or placed suddenly at any angle which threw a heavy strain on their planes. A machine for exhibition flying needs to be constructed specially; but this was not realised till accidents had taught their lesson.

It is a regrettable fact, one which emerges directly from a study of aeroplane accidents, that many of them might have been avoided had men been content to follow warily in the footsteps of the pioneers, and not run heavy risks till they themselves, and the machines they controlled, had been prepared, by a long period of steady flying, to meet such greater dangers. The first men who flew realised fully the risks they ran. But when flying became more general, and men found machines ready to their hands, machines which it was a simple matter to learn to fly, this early spirit of caution was forsaken, and feats were attempted which brought fatalities in their train, and which seemed to emphasise the risks of aviation, and did it the very bad service that they fixed in the public mind a notion of its dangers, and prevented men from coming forward to take up flying as a sport.

CHAPTER VIII

FACTORS THAT MAKE FOR SAFETY

It has been calculated that nearly half the aeroplane disasters of the early days were due to a structural weakness in machines, or to mistakes either in their design, or in such details as the position, shape, and size of their surfaces. To-day, thanks to science, and to the growing skill and experience of aeroplane designers and constructors, this risk of the collapse of a machine in the air, or of its failure to respond to its controls at some critical moment through an error in design, has been to a large extent eliminated. That such risks should be eliminated wholly is, as yet, too much to expect.

One of the factors making for safety has been the steady growth in the general efficiency of aircraft: in the curve of their wings which, as a result largely of scientific research, has been made to yield a greater lift for a given surface and to offer a minimum of resistance to their passage through the air; in the power and reliability of their engines; in the efficiency of their propellers; and in the shaping of the fusilage of a machine, and in the placing and "stream-lining" of such parts as meet the air, so as to reduce the head resistance which is encountered at high speeds. Such gains in efficiency, which give constructors more latitude in the placing of weight and strength where experience show they are needed, have gone far to produce an airworthy machine. In the old days, when machines were inefficient, a few revolutions more or less per minute in the running of an engine meant all the difference between an ascent and merely passing along the ground. But nowadays, through the all-round increase in efficiency that has been obtained, a machine will still fly upon its course without losing altitude, and respond to its controls, even should the number of revolutions per minute of its engine be reduced considerably.

When given a greater efficiency in lifting surfaces and power-plants--and profiting also from the lessons that had been learnt in the piloting of machines--constructors were able to devote their attention, and to do so with certainty instead of in a haphazard way, to the provision of factors of safety when a craft was in flight. With a machine of any given type, if driven through the air at a certain speed, it is possible to estimate with accuracy what the normal strains will be to which it is subjected. But even if such data are obtained, and the machine given the strength indicated, this factor of safety is insufficient. It is not so much the normal strains, as those which are abnormal, that must be guarded against in flight. A high-speed machine, if piloted on a day when the air is turbulent, may be subjected to extraordinarily heavy strains; rising many feet in the air one moment, falling again the next, and being met suddenly by vicious gusts of wind--in much the same way that a fast-moving ship, when fighting its way through a rough sea, is beaten and buffeted by the waves. Air waves have not of course the weight, when they deliver a blow, that lies behind a mass of water; but that these wind-waves attain sometimes an abnormal speed, and have a tremendous power of destruction, is shown in the havoc that is caused by hurricanes.

It seems astonishing to many people that such a frail machine as the aeroplane, with its outspread wings containing nothing stronger often than wooden spars and ribs, covered by a cotton fabric, should be capable of being driven through the air at such a speed, say, as 100 miles an hour, encountering not only the pressure of the air, but resisting also the fluctuations to which it may be subjected. But, underlying the lightness and apparent frailty of such a wing, when one sees it in the workshop in its skeleton form, before it has been clothed in fabric, there is a skill in construction, and an experience in the choice, selection, and working of woods, that produces a structure which, for all its fragile appearance, is amazingly strong. And the same applies, nowadays, to all the other parts of an aeroplane. That it should have taken years to gain such strength, and to reduce so largely the risk of breakage, is not in itself surprising. Men had to devise new methods in construction--always with the knowledge that weight must be saved--and to create new factors of safety, before they could build an airworthy craft.

To-day, when a man flies, he need have no lurking fear, as had the pioneers, that his craft may break in the air. Even when it is driven through a gale, plunging in the rushes of the wind, yet held straining to its task by the power of its motor, the modern aeroplane can be relied upon; and not in one detail of its construction, but in every part. Experience, the researches of science, and the growing skill with which aircraft are built, stand between the airman and many of his previous dangers. The aeroplane to-day, one of the structural triumphs of the world in its lightness and its strength, has a factor of safety which is sufficient to meet, and to withstand, not merely ordinary strains, but any such abnormal stresses as it may encounter--and which may be many times greater than the strains of normal flight.

The aviator knows also that his engine, as it gives him power to combat successfully his treacherous enemy, the wind, represents the fruit of many tests and of many failures, and of the spending of hundreds of thousands of pounds. Many of its defects have revealed themselves, and been rectified; it is no longer light where it should have weight of metal, nor weak where it should be strong. So far as any piece of mechanism can be made reliable, consisting as it does of a large number of delicate parts, operating at high speed, the aeroplane motor has been made reliable. But, so long as one motor is used, there must always, as we have said, remain a risk of breakdown. It is for this reason that, thanks largely to the stimulus of the war--which has created a practical demand for such machines--aeroplanes are now being built, and flown with success, which are fitted with duplicate motors. With such machines, which give us a first insight as to the aircraft of the future, engine failure begins to lose its perils--particularly in regard to war. More than once during the great campaign, when flying a single-engine machine, an aviator has found his motor fail him, and has been obliged to land on hostile soil; with the result that he has been made prisoner. But with dual-engine machines it has been found that, when one motor has failed mechanically, or has been put out of action by shrapnel, the remaining unit has been sufficient--though the machine has flown naturally at a reduced rate--to enable the pilot to regain his own lines.

In peace flying, too, as well as in war, the multiple-engined aeroplane brings a new factor of safety. If one of his motors fails, and he is over country which offers no suitable landing-place, the pilot with a duplicate power-plant need not be concerned. His remaining unit or units will carry him on. There are problems with duplicate engines which remain to be solved--problems of a technical nature--which involve general efficiency, transmission gear, and the number and the placing of propellers; but already, though this new stride in aviation is in its earliest infancy, results that are most promising have been obtained.

To those who study aviation, and have done so constantly, say from the year 1909, one of the most striking signs of progress lies in the fact that, though unable at first to fly even in the lightest winds, the aviator of to-day will fight successfully against a 60 miles-an-hour wind, and will do battle if need be, once he is well aloft, with a gale which has a velocity of 90 miles an hour. He will ascend indeed, and fly, in any wind that permits him to take his machine from the ground into the air, or which the motor of his craft will allow it to make headway against. And here, though machines are still experimental, there is removed at one stroke the earliest and the most positive objection of those who criticised a man's power to fly. When the first aeroplanes flew the sceptics said: "You have still to conquer the wind, and that you will never do. Aeroplanes will be built to fly only in favourable weather, and this will limit their use so greatly that they will have no significance." But to-day the aviator has ceased, one might almost say, to be checked or hampered by the wind. If the need is urgent, as it often is in war, then it will be nothing less than a gale that will keep a pilot to the ground, provided he has a sufficiently powerful machine, and a suitable ground from which to rise--and granted also that he has no long distance to fly. Wind-flying resolves itself into a question of having ample engine-power, of being able to launch a machine without accident, and get it to earth again without mishap; and of being able to make a reasonable headway against the wind when once aloft; and these difficulties should solve themselves, as larger and heavier machines are built.

Apart from the growing skill of the aviator, which has been bought dearly, science can now give him a machine, when he is in a wind, that needs no exhausting effort to hold it in flight. Craft are built, as a matter of certainty and routine, which have an automatic stability. Science has made it possible indeed, by a mere shaping and placing of surfaces, and without the aid of mechanical devices, to give an aeroplane such a natural and inherent stability that, when it is assailed by wind gusts in flight, it will exercise itself an adequate correcting influence. To understand what this means it should be realised that, when such a machine is in flight say in war on a strategical reconnaissance, and carries pilot and passenger, the former can take it to a suitable altitude and then set and lock his controls, and afterwards devote his time, in common with that of his passenger, to the making of observations or the writing of notes. The machine meanwhile flies itself, adapting itself automatically to all the differences of wind pressure which, if it had not this natural stability, would need a constant action of the pilot to overcome. All he need do is to maintain it on its course by an occasional movement of the rudder. With such a machine, even on a day when there is a rough and gusty wind, it is possible for an airman to fly for hours without fatigue; whereas with a machine which is not automatically stable, and needs a ceaseless operation of its controls, the physical exhaustion of a pilot, after hours of flight, is very severe.

So, already, one sees these factors of safety emerge and take their place. There is no longer a grave peril of machines breaking in the air; there need be no longer, with duplicate power-plants, the constant risk of engine failing; while that implacable and treacherous foe, the wind, is being robbed daily of its perils.

CHAPTER IX

A STUDY OF THE METHODS OF GREAT PILOTS

The masters of flying, and this is a fact the novice should ponder well, have been conspicuous almost invariably for their prudence. No matter how great has been their personal skill, they have never lost their respect for the air; and this is why so many of the great flyers, after running the heaviest of risks in their pioneer work, have managed to escape with their lives. What patience and sound judgment can accomplish, when pitted even against such dangers as must be faced by an experimenter when he seeks to fly, is shown by an incident from the early career of the Wright brothers. With one of their gliders, a necessarily frail machine, and in tests made when they were both complete novices, they managed to make nearly 1000 glides; and not once in all those flights, during which they were learning the rudiments of balance and control, did they have a mishap which damaged at all seriously their machine.

These two brothers, Wilbur and Orville, offer to the student of flying, apart from the historical interest which is attached to their work, a temperamental study of the greatest interest. Wilbur, who was grave, judicial--a man of infinite patience and with an exceptional power of lucid thinking--found in his brother and co-worker, Orville, a disposition just such as was necessary to strengthen and support him in his great research; a disposition more vivacious and more enthusiastic than his, and one which acted as a balance to his own gravity. The method of these brothers in first attacking a mass of data, most of it contradictory--and a large amount of it of little intrinsic value--and then framing their own research on lines which they discussed and studied with methodical care, forms a model of sound judgment for workers in any complex field. Their kite experiments, their gliders, their refusal to hasten their steps unduly in the fitting of an engine to their machine, reveal again their discretion, and that judgment which never failed them. Perseveringly and unswervingly, exhibiting doggedness without obstinacy, and with their work illuminated always by the highest intelligence, they moved surely from stage to stage; and at last, when they fitted a motor to their machine, such was their knowledge of the air, and of the control of their craft when in flight, that they were able to make this crucial step, from a glider to a machine driven by power, without any breakage of their apparatus or injury to themselves.

The same self-control marked them when, having demonstrated that men can ascend in a power-driven machine, and steer such a craft at will, they dismantled their apparatus and commenced their negotiations with foreign Governments. Wilbur Wright, too, when he came to France to give his first public demonstrations, provided by his methods a model for aviators, either present or future. He resisted all temptations to make injudicious flights. If he considered the weather conditions at all unsuitable he said that he would not ascend, no matter who might have come to see him fly, and that settled the question once and for all. He was deaf to all pleadings, to all proffered advice. When conditions were perfectly suitable, and then only, would he have his craft brought from its shed.

The same meticulous care, in every flight he made, marked his preparation of his machine. Motor, controls, propeller-gearing, every vital part, received its due attention; and this attention was never relaxed, no matter how frequently he flew, nor how great was his success. An observer of one of his early flights at Le Mans has given us an impression that is typical of this unremitting care. There was a question of some small adjustment that Wilbur had instructed should be made to the machine. When the time came to fly, and he was in the driving-seat waiting for the motor to be started, he called a question as to whether this detail had been attended to. He was assured it had. But this was not enough for Wilbur Wright. Climbing from his seat and walking round the biplane, he made a careful examination for himself, and then returned quietly to the front of the machine. People who came to see him fly, and expected some picturesque hero, leaping lightly into his machine and sweeping through the air, found that reality disappointed them. This quiet, unassuming man, who slept in his shed near his aeroplane, and took his meals there also, refused to be feted or made a fuss of; while his deliberation in regard to every flight, and his indifference to the wishes or convenience of those who were watching him, drove nearly frantic some of those influential people who, coming in motor-cars and with a patronising spirit, thought the aviator might be treated rather as a superior mountebank, who would be only too glad to come out and fly when a distinguished guest arrived.

M. Louis Bleriot, whose name was next to become world-famous, after that of the Wrights, and who owed his distinction to his crossing of the English Channel by air, revealed in his character determination and courage, and imagination as well. And yet allied to these qualities--and here lay his temperamental strength--he had a spirit of quiet calculation and an eery considerable shrewdness. He knew, and was not afraid of showing that he knew, the full value of caution. And yet on occasion also--as in the cross-Channel flight--he was ready to put everything to the test, and to take promptly and with full knowledge the heaviest of risks. The motor in his cross-Channel monoplane was an experimental one of low power, air-cooled, and prone to over-heat and lose power after only a short period of running. To cross the Channel, even under the most favourable circumstances, he knew this engine must run without breakdown for thirty-five or forty minutes. This it had not done--at any rate in the air--before. There was a strong probability--and Bleriot knew this better than anyone else--that the motor would fail before he reached the English shore, and that he would have to glide down into the sea. It was arranged that a torpedo-boat-destroyer should follow him, and this afforded an element of safety. But Bleriot guessed--as was actually the case--that he would outdistance this vessel in his flight, and soon be lost to the view of those upon it. And he did not deceive himself as to what might happen, if his engine stopped and he fell into the water. His monoplane, as it lay on the surface of the water, would, he knew, prove a very difficult object to locate by any vessel searching for it; while it was so frail that it would not withstand for long the buffeting of the waves. He carried an air-bag fixed inside the fusilage, it is true; but, in spite of this precaution, Bleriot knew he ran a very grave peril of being drowned. There was, on the morning of his flight, another disturbing factor to be reckoned with. The wind, calm enough when he first determined to start, began very quickly to rise; and by the time he had motored from Calais to the spot where his aeroplane lay, and the machine itself was ready for flight, the wind out to sea was so strong that the waves had become white-capped. But Bleriot, aware of the value at such moments of decision, had made up his mind. He knew that, if his engine only served him, his flight would be quickly made. And so he reckoned that, even though the wind was rising, he would be able to complete his journey before it had become high enough seriously to inconvenience him; and in this calculation, as events proved, he was right. His motor did its work; and, though the wind tossed his machine dangerously when he came near the cliffs of the English coast, he succeeded in making a landing and in winning the ?000 prize.

M. Hubert Latham, Bleriot's competitor in the cross-Channel flight, had that peculiar outlook on life, with its blend of positive and negative--puzzling often to its owner as well as to the onlooker--that is called, for the sake of calling it something, the artistic temperament. He was impulsive, yet impassive often to a disconcerting extent: extremely sensitive and reserved as a rule, yet on occasion almost boyishly frank and communicative. He lacked entirely ordinary shrewdness, or everyday commonsense. He was a man of a deeply romantic temperament, and this inclined him towards aviation and the conquest of the air; while in actual piloting he had such a quickness and delicacy of touch, and such a sure and instinctive judgment of distance and of speed, that he was undoubtedly a born aviator--one of, if not the, finest the world has seen. That he did not attain greater success, from a practical point of view, was due to the fact that he was without the level-headedness and the business ability which characterised others of the pioneers. When he was in flight in his Antoinette--Latham flew that machine and no other--he was a supreme artist. His machine was beautiful, and his handling of it was beautiful.

M. Henri Farman, beyond question, of course, another of the great pioneers, is a man of imagination and of a highly nervous temperament, yet possessing at the same time a very pronounced vein of caution. No success has for an instant caused him to lose his head. At Rheims, in 1909, when he had created a world's record by flying for more than three hours without alighting, those who hastened to congratulate him, after his descent, found him absolutely normal and unmoved. Washing his hands at a little basin in the corner of the shed, he discussed very quietly and yet interestedly, and entirely without any affectation of nonchalance, the details of his flight and the behaviour of his motor. His attitude was that the flight was something, yet not a great deal, and that very much more remained to be done; a perfectly right and proper attitude, one which was just as it should be, yet one encountered very rarely under such circumstances--human nature being what it is.

Farman's patience, his perseverance, were in the very early days what gave him his first success. With the biplane the Voisins built him, for example, nothing but his own determination, and his ceaseless work upon his engine, enabled him to do more with this type of machine than others had done.

As the aeroplane increased in efficiency, and in the reliability of its engine, and was used in cross-country journeys, there came an era of flying contests, in which large prizes were offered, and in which airmen passed between cities and across frontiers, and traversed in their voyages the greater part of Europe. In the making of these flights, which needed an exceptional determination and skill, allied also to a perfect bodily fitness, there came into prominence certain aviators whose precision in their daily flights, passing across country with the speed and regularity of express trains, won admiration throughout the world. Prominent among these champions was the French naval officer, Lieut. J. Conneau, who adopted in his contests the flying name of "Beaumont." His success and his exactitude, when piloting a Bleriot monoplane for long distances above unknown country, guiding himself by map and compass, gave the public an indication, for the first time, of what might be accomplished by an expert airman when flying a reliable machine. Lieut. Conneau's success, winning as he did several of the great contests one after another, and the absence of error in his flying from stage to stage, and his accurate landings upon strange and often badly-surfaced aerodromes, should provide for the novice in aviation--when the secret of this success is understood--an object-lesson that is of value.

This quiet, efficient airman, and his methods in making himself so competent, afford indeed an interesting study. Here was one who, suited already by temperament for the tasks he undertook, trained himself with such care, with such patience, that he attained as nearly to the ideal as is possible for living man. When he had asked for, and obtained, permission from the Minister of Marine to study aviation in all its aspects, he began his task in a spirit that was admirable. "I was convinced," he wrote afterwards, "that a perfect knowledge of machines and motors was necessary before one could use them." For nearly a year, on leaving the sea, he worked to obtain a certificate as a practical engineer. This gained, he went through a period of motor-cycling and motor-car driving, varied by flights in captive balloons and free balloons, and afterwards in airships. Following this he obtained leave to stay for a time at Argenteuil, and enter the works of the builders of the Gnome motor. Here he lived the life of a mechanic, and learned to understand completely the operation of this famous engine, which he was destined to drive afterwards in his great flights.

Presently he went to Pau, in order to obtain his certificate as an aeroplane pilot. At first, taking his turn with a number of other pupils, he could only get a few minutes at a time in a machine. But being a keen observer he found that, by listening to the instructors, and watching the flights made, he could pick up useful information without being in the air; and this led him to the observation that "to learn to fly quickly, one must begin by staying on the ground."

He secured in due course his certificate of proficiency, astonishing the instructors by his skill and sureness in the handling of his machine. Then followed what might be called an apprenticeship to cross-country flying. He made constant flights in all weathers, flying for instance from Pau to Paris, and studying closely not only the piloting of his machine and the aerial conditions he encountered, but also the art of using a map and compass, and in finding a path without deviation from point to point. Improving daily in confidence and skill, and learning practically all there was to be learned as to the handling of a Gnome-engined Bleriot, he was able soon to fly under weather conditions which would have seemed hopeless to a pilot of less experience; while engine failure and other troubles, which overtook him frequently on these long flights, taught him to alight without damaging his machine on the most unpromising ground.

Now, feeling himself at last competent, he obtained permission to figure on the retired list, so that he might take part in the aviation races which were then being organised. Of these great contests Lieut. Conneau won three in succession--the Paris-Rome Race, in which he flew 928 miles in 21 hours 10 minutes; the European Circuit, in which he flew 1,060 miles in a total flying time of 24 hours 18 minutes; and the Circuit of Britain, in which he flew 1,006 miles in 22 hours 26 minutes. Lieut. Conneau's success, which appeared extraordinary, and his skill in finding his way across country, which seemed abnormal, were due as a matter of fact to his assiduous preparation, and to a temperament which, even under the heavy strains of constant flying, saved him from errors of judgment or ill-advised decisions. His temperament was, indeed, ideal for a racing airman. He was quiet and collected, with a natural tendency to resist excitement or confusion. His physique was admirable, and he had that elasticity of strength, both in body and nerve, which are invaluable to a pilot when on long flights. Also, and this was of importance, Lieut. Conneau had a natural cheerfulness of disposition which carried him without irritation or despondency through those ordeals of weather, and of mechanical breakdowns and delays, which are inseparable from such contests as those in which he was engaged.

A contrast to Lieut. Conneau, both in temperament and method, was his rival Jules Vedrines--the aviator who, notably in the Circuit of Britain, flew doggedly against Lieut. Conneau from stage to stage. Vedrines, who had not had the advantages in tuition that had been enjoyed by Lieut. Conneau, nor his grounding in technique, was nevertheless a born aviator; a man of a natural and exceptional skill. In energy, courage, and determination he was unexcelled; but such qualities, though of extreme value in a long and trying contest, were marred by an impetuosity and an excitability which Vedrines could not master, and which more than once cost him dear. He had not, besides, as was shown in the Circuit of Britain, that skill in steering by map and compass which aided Lieut. Conneau so greatly in all his flying.

A personality of unusual interest was that of the late Mr. S. F. Cody--a man of a great though untutored imagination, and of an extraordinary and ceaseless energy. A big man, and one whom it might be thought would have been clumsy in the handling of an aeroplane, he piloted the biplanes of his own construction with a remarkable skill. He flew no other, of course, and this was greatly to his advantage in actual manipulation. The great pilots who have excelled--one may instance again Lieut. Conneau--have concentrated their attention as a rule on one type of machine, learning all there is to be learned about this particular craft, and being prepared in consequence, through their knowledge both of its capacities and weaknesses, for any contingency that may arise in flight. Another instance of such specialisation was provided by Mr. Gustave Hamel. M. Bleriot--an admirable judge in this respect--singled out Mr. Hamel, while this young man was learning to fly in France, as an aviator of quite unusual promise; and his prediction was, of course, more than fulfilled. Devoting himself exclusively to the monoplane, Mr. Hamel became a pilot whose perfection of control, very wonderful to witness, was marked strongly by his own individuality. He had beautiful "hands"--a precision and delicacy on the controls which marked his flying from that of all others; while his judgment of speed and distance, which was remarkable, represented natural abilities which had been improved and strengthened by his constant flying.

CHAPTER X

CROSS-COUNTRY FLYING

When a pupil has finished his flying school tests, and has received his certificate from the Royal Aero Club, he is in a stage of proficiency which means that he has learned to control an aeroplane when above an aerodrome and in conditions that are favourable, and that he may be relied on to make no elementary mistakes. But as to cross-country flying, with its greater hazards, he is still a novice, with everything to learn. And so it is to flights from point to point, generally between neighbouring aerodromes, that he next devotes himself.

Aviators have been commiserated with, often, on what is thought to be the monotony of a cross-country flight. The pilot, raised to a lonely height above the earth, is pictured sitting more or less inertly in his seat, with nothing to do but retain his control on the levers, and look out occasionally so as to keep upon his course. But the beginner, when he first attempts cross-country flying, will have an impression not of inactivity, but of the necessity to be constantly on the alert. He will be engrossed completely by the manipulation of his machine, with no time to sit in idle speculation, or to analyse his feelings as the country passes away below.

When preliminaries on the ground have been gone through, and the pilot is in the air, there will first be a need to gain a height of several thousand feet. Altitude is essential in cross-country flying. The higher a pilot flies, within reason and having regard to the state of the atmosphere, the better chance will he have of making a safe landing, should his motor fail suddenly and force him to descend. So the first concern is climbing--and in doing so the pilot must remember the teachings of his instructor, and not force his craft on too steep or rapid an ascent. He may prefer, in his early flights, to remain above the aerodrome while he is gaining altitude, watching his height recorder from moment to moment so as to note his progress upward. He will be occupied also with his engine, listening to its rhythm of sound, and keeping an eye on the indicator that tells him how many revolutions per minute the motor is actually making, and which will warn him at once should it begin to fail.

Granted his motor is running well, a pilot should soon gain altitude. Then, assuming the air is clear--as it should be on his early flights--he will note some landmark, away on the line of his flight, and set off across country towards it. Fixed conveniently in front of him will be a map, of a kind devised specially for the use of aviators. A pilot's view, as he flies high above the ground, is bird-like. Landmarks fail to attract his attention, at this altitude, which would be clearly seen if he were on the ground. Hills, for example, unless they are high, are so dwarfed as he looks down on them that they scarcely catch his eye. What is done, by the designer of air maps, is to accentuate such details of a landscape as will prove conspicuous when seen from above. A river, or an expanse of water, is clearly seen; so also are railways and main roads; while factory chimneys, and large buildings which stand alone, may be identified from a distance when a pilot is in flight. So on an airman's map, made to stand out by various colourings in a way that catches the eye, are railways, roads, rivers, lakes and woods, with here and there a factory chimney or a church, should these be in a position rendering them visible easily from the air. That such maps should be bold in their design, and free from a mass of small details, is very necessary when it is remembered that the aviator, passing through the air at high speeds, has no time for a leisurely inspection of his map.

With a good map, and aided when necessary by the compass that is placed in a position so that he can see it readily, a pilot has no difficulty as a rule, once he has acquired the facility that comes with practice, in steering accurately from point to point, even when on a long flight. On a favourable day, when the land below is clearly visible, he will glance ahead, or to one side, and after observing some landmark, look on his map to identify the position he has just seen. Under such conditions steering is easy, and the compass plays a subsidiary part. But it may happen that, while he is on a long flight and at a considerable altitude, the earth below may be obscured by clouds, or a low-lying mist, and all landmarks vanish from his view. Sometimes too, he may find himself flying through mist and cloud, with all signs of the earth gone from below. Whereupon, robbed for awhile of any direct guidance, he must fly by aid of his map and compass, holding his machine on its compass course, and noting carefully the needle of his height-recorder, so that he is sure of maintaining altitude. A risk exists under such conditions, when there is no visible object by which to judge a course, that an airman may make leeway, unconsciously, under the pressure of a side-wind; and so he must be ready to note carefully, immediately that a view of the earth is vouchsafed him, whether he has actually been making leeway, either to one hand or the other, even while the bow of his machine has been held on its compass course. There is a risk also, when a pilot is flying in fog or at night, that, having no visible horizon from which to gauge the inclination of his craft, it may assume gradually some abnormal angle, without his own sensations telling him what is taking place. The craft may, for the sake of illustration, incline sideways, imperceptibly to the pilot, till it begins to side-slip. But science can meet this danger by providing inclinometers, fitted within the hull so that the aviator can see them easily; and by means of these instruments, which are illuminated at night, it is possible for a pilot to tell, merely by a glance, at what angle his machine is moving forward through the air--whether it is up or down at the bow, or whether its position laterally is normal.

The beginner, on his first cross-country flight, need not be troubled by such intricacies. He is flying, one assumes, on a fine day, with the land spread clearly below him. So as he moves through the air, listening always to the hum of his motor, he need have no fear, granted that his observation is ordinarily keen, of losing his way.

Naturally, being a novice, he will feel the responsibility of his position. His eyes will rove constantly from one instrument to another; as indeed, from habit, do those of a practised flyer. He will glance at the height recorder; then at the engine revolution indicator; then at the dial which tells him what his speed is relative to the air. There is a dial, also, showing the pressure in his petrol-tank; while there will be a clock on his dashboard at which he will glance occasionally, after he has marked some position away on the land below, so as to determine what progress he is making from the point of view of time.

Besides these preoccupations, and the ceaseless even if almost unconscious attention that he must pay to his engine, there is the need to bear constantly in his mind's eye the lie of the land. Should his motor fail suddenly, or something happen which necessitates an immediate descent, it is imperative that he should be able, without delay, to choose from the ground that is visible below him some field or open space that will provide a safe landing-point. And this is easier said than done. The earth, when viewed by a airman who looks down almost directly upon it, is apt to be deceptive as to its contour. A field that is selected say, from a height of several thousand feet, may not prove--as the aviator nears it in his glide--to be at all the haven he imagined it. More than once, seeking to alight on a field which appeared to him, as he was high above it, to be level as a billiard table, a pilot has found, when it is too late, that the ground has sloped so steeply that his machine, after landing, has run on downhill and ended by crashing into a fence or ditch.

It is very necessary for an airman to learn to judge, by its appearance, the difference between an expanse, say, of pasture land, or a field which is in green corn or standing hay. It has happened often that a pilot, descending after engine failure towards what he has reckoned a grass field, has discovered--when too low to change his landing-point--that his pasture land is actually a field of green corn; and a landing under such conditions, with the corn binding on the running-gear of the machine, may end in the aircraft coming to an abrupt halt, and then pitching forward on its nose; with a broken propeller and perhaps other damages in consequence.

In choosing a landing ground, as in other problems that face the novice in cross-country flying, experience will prove his safeguard. He will learn for instance that cattle or sheep, if they can be discerned below in a field, go to show that this field is one of pasture and not of crops. If no cattle are to be seen in a field, and the aviator is doubtful about it, and yet if it happens to be the only suitable one he can locate, then he may look closely at the gateway which leads into the field. If, in this gateway, he can detect such scars or markings on the ground as are caused by the feet of cattle as they walk daily in and out, he may feel satisfied the field is one of pasture.

When cattle or sheep are seen standing in a field so that they face in the same direction, this may suggest either the existence of a slope, or the presence of a strong ground wind; while a stream or brook at the edge of a stretch of open land, or a belt of woods, may suggest a sloping of the ground.

It is amusing for a pilot--or it was so, rather, in the days when few aeroplanes were in existence--to note the astonishment which his descent, made quite unexpectedly perhaps in some quiet and rural country, will occasion amongst the inhabitants. Sometimes, under the stress of such an excitement, people appear to lose for the time being their power of coherent speech. A pilot in a cross-country contest, not being sure whether he was on his right course, decided to make a landing and ask his way. He noticed, after a while, the figure of a man in a field below. Planing down, and alighting in the field, he shouted questions to this man, switching his engine off and on, while he did so, in order that his words, and those of the other, might be audible. But the man in the field, demoralised by the advent of this being from the air, and gazing at him and his machine with an expression of blank amazement, was unequal to the task of giving even the simplest directions. He waved his arms, it is true, but no words that could be understood issued from his lips. The pilot repeated his questions, but it was no good. The man waved and mouthed, and rolled his eyes, but when he tried to speak intelligibly he could not. So the aviator, loath to waste further time, accelerated his engine again and continued his flight.

As a contrast to this, there was the experience of a pilot who, after a long flight from England to the Continent, landed at length near a small village. In the next field to that in which he alighted there was a labourer, digging patiently. The aviator expected that this man would fling down his spade in excitement, and run wildly towards the aeroplane. But such was not the case. This labourer, a marvel of placidity and unconcern, merely raised his head slowly and looked across at the aircraft, and then went on with his digging.

In his first cross-country flights, being concerned chiefly as to the manipulation of his machine, and having so many things to think of, the novice may feel tired after even a short journey by air. His chief sensation, as he switches off his engine to descend towards the aerodrome he sees below him, will be one of relief that he has escaped engine failure, and that he has been able to find his way from point to point. The joy of flight, of passing swiftly thousands of feet above the earth, will have made but a small impression upon him--at any rate consciously. It will not be until the handling of his machine becomes less laborious, and he has time to accustom himself to his unique view-point, and the strangeness and beauty of the scene below him, that the novice will realise some of the fascinations of aerial travel; fascinations that it is difficult to describe. The sensation of having thrown off the bonds of earth-bound folk; of soaring above the noise and dust of highways; of being free from the obstructions of traffic; of sweeping forward smoothly, swiftly, and serenely--the land stretching below in an ever-changing panorama, with the drone of the motor in one's ears, and a wine-like exhilaration in the rush of the air: these, and others more obscure, are among the sensations of cross-country flying.

CHAPTER XI

AVIATION AS A PROFESSION

Young men, and parents on their behalf, are seeking always some profession which will yield an adequate return for the enthusiasm which youth lavishes upon it. Too often, though, at any rate in the past, this search for a man's work in life has been narrowed into ruts; conducted on certain set lines which, though they have found employment for the beginner, have given him no scope for that enthusiasm with which he will attack the first tasks presented him. Aviation, till the coming of the war, was looked at askance by parents who had sons on their hands. Apart from the risks of flying, which appeared to them ceaseless and terrible, the actual industry of building aeroplanes, regarded as an industry, seemed so haphazard and objectless an affair--so much like playing at work--that they discouraged any wish that a youth might show to enter it. Many people, these people of intelligence, regarded the building and flying of aeroplanes as being no more than a passing phase, and a regrettable one, which it was hoped men would soon abandon, and turn their attention to tasks more serious and profitable. But that was before aircraft had proved their value as instruments of war. Now it is known that aeroplanes have the power, granted they are supplied in sufficient numbers, of altering the tenor of a great campaign, both by land and sea; and that in any future war of nations, should one come, a battle between the hostile flying fleets, fought to determine the command of the air, will determine also, to a very large extent, the fortunes of armies on the land and navies on the sea. It is clear indeed that, for any great nation that strives to maintain its place, a powerful air fleet has become a necessity; while for Britain, an island no longer from the military point of view, seeing that we must face seriously the question of invasions by air, there is a vital need to strive for command of the air, as we now hold command of the sea.

The building up of our air fleet will be an arduous task, needing men, money, and time; but without it we cannot be secure. Therefore the work must be faced, the men and the money forthcoming. Aviation, as an industry, must prepare for years of strenuous work. A great air service must be created. Machines must be designed and built in thousands instead of hundreds, and men trained to fly them. Nor is this all. The aeroplane, though it has such significance as a weapon of war, is destined primarily and eventually to be an instrument of peace; a machine for the transport by air of passengers, mails, and goods, at speeds greater than will be feasible by land or water; and a craft also for the use of travellers and tourists, enabling them to make such journeys, with ease and pleasure, as will again prove impossible by land or sea. So aviation has two immense tasks ahead of it, instead of one. Not only must it create, by years of patient and determined effort, a flying service which will command the air, but craft must be designed and built also for the mail, goods, and passenger-carrying services, and to meet the needs of the aerial tourist.

This new task that has been given to men, that of designing, building, and piloting aircraft, is still on the eve of its expansion. The opportunities it offers to young men--to men whose minds are quick to grasp a new idea and who have the powers of initiative and decision--are almost boundless. Flying will, as it develops, revolutionise the world's system of transport; while the developments even of the immediate future promise to be so great, and so important, that it is not easy to visualise them. But this at least is clear: now is the time for newcomers to enter the world of flight. Aviation needs men, is calling aloud for men; and they are needed for many kinds of work. First, of course, should be placed the flying services, naval and military, to join which during the war men have come forward so admirably. But it will need, in the expansion that must follow this campaign, a steady and a ceaseless growth in numbers, not only of the men who handle machines in flight, but of those who serve the squadrons by their work on land, and who build up the organisation which is vital to success.

For skilled aviators, other than those who join the services, there is scope for remunerative work. A constant demand exists for men who will test and fly in their trials the new machines that are built by manufacturers; for men who will fly, in public exhibitions, the craft that are used at the various aerodromes; and for men who will qualify as instructors, and join the flying schools which are already in existence, or in process of formation. In countries oversea, too, there is the definite promise that aircraft will be needed, not only for survey work over wide tracts of land, and for maintaining communication and bearing mails over districts where land travel is difficult, but also for exploration; and this again means that pilots will be required. New aerodromes must come into existence also; not only to act as alighting points for touring craft, but to provide grounds for the training of pupils; and at these aerodromes pilots will be needed.

Of other opportunities, apart from the piloting of aircraft, there are many--though it is desirable for a man to learn to fly, and obtain his certificate of proficiency, even if afterwards he does not intend continuing as a pilot. The practical experience he gains, while learning actually to handle an aircraft in flight, will prove extremely useful to him subsequently, even though the task he undertakes is one that keeps him on the ground. He may qualify, for instance, for a post in a aeroplane factory as a designer or draughtsman; or he may specialise in aero-motors, and seek a post in the engine-shops. At the aerodromes, too, there are openings which present themselves; as, for example, in the management of a flying school.

It has been shown that the public will go in thousands to see sporting contests with aeroplanes, and here is another field for organisation and effort; while there is a constant demand for men of ability in the executive departments of firms which are established already in the industry, and are expanding steadily, or in those which are now being formed, or are joining aviation from day to day.

The industry is at last on a footing that is practical and sound. It presents a new field for effort, and one that is unexploited; while for the man who enters it--and this should be the attraction for youth--there are occupations as fascinating as one's imagination could depict. But one thing must be understood clearly. Flying is, of exact sciences, surely the most exact. The man who is only half-trained, who is more or less slovenly in his work, who will not bend his whole energies to his task, will find no place in this new industry. A young man is wasting his time, if, after deciding to enter aviation, he acquires knowledge that is no more than haphazard. He who contemplates aviation as a profession must set himself the task of learning all there is to be learned, and in the right way.

Individual opportunities and circumstances will, necessarily, play so large a part in the steps taken by a young man--or by his parents on his behalf--to launch him on a career in aviation that it is impossible, here, to do more than generalise. Certainly, as we have said, it is an excellent preliminary to learn to fly; and it may be stated also that it is now possible to place, with aviation companies of repute, premium pupils who will undergo instruction extending over a period of three years. A youth may, also, gain his knowledge of the industry by becoming an indentured apprentice.

One may say, as a conclusion to this chapter, that a great, new, and potential industry is springing up in our midst, one which will prove to be equally if not more important and far-reaching than the British shipbuilding industry, and one which will employ thousands of skilled engineers and artisans. Ships are limited to one element, the water, which has very definite boundaries. Aircraft, too, are limited to one element, the air; but this element has no boundaries so far as the earth is concerned, and aircraft will be navigable to any and every part of the globe.

CHAPTER XII

THE FUTURE OF FLIGHT

It is a hopeful augury, to those concerned with aviation, that public interest in flying should not only be keen, but should be growing. In the early days, even when aeroplanes were so great a novelty, it was difficult to induce people in any numbers to witness a flying display. The first meetings, though they were organised with enthusiasm, ended as a rule with a heavy financial loss; and this fact of course, when it became known, had a discouraging influence on those who might, had these early meetings proved a success, have been willing to finance aerodromes and the building of machines. But as it was, business men, who are quick to form conclusions, said that people would never be induced to pay to see aeroplanes fly. But here they failed to reckon with the fact that, though public interest in flying has been of very slow growth, yet at the same time it has been a steady and continuous growth. From month to month, and from year to year, as aeroplane constructors and pilots have continued at their tasks, overcoming technical difficulties and personal risks, the interest of ordinary people has grown perceptibly. Even before the war--which has done so much to focus attention on flying--the attitude of scepticism and apathy had been greatly changed. When the London Aerodrome at Hendon was established, there were shrewd men in the city, men who are ordinarily very sound in their conclusions, who declared the public would never go there in appreciable numbers. How wrong they were, how little they gauged the change that was taking place in the public mind, is shown by the fact that, on a popular day at this aerodrome, as many as 60,000 people have paid for admission.

In the immediate future, as in the immediate past, aviation will be concerned largely with the building of naval and military craft. This will, so to say, be the foundation of its development in other directions. War for instance, notably in the fitting of craft with duplicate power-plants, will provide data that is invaluable in the building of commercial craft, and in machines also for the use of the tourist. In aerial touring there lies an important field for the development of aircraft--one which may serve to bridge the gap between a relatively small, purely pleasure-type machine, and a craft which has utility in the fields of commerce. The motor-car provides an enjoyable means of travelling from place to place; but in the aeroplane, once it is airworthy, reliable, and comfortable, the tourist has a vehicle which is distinctly more pleasurable and exhilarating. The day was dawning before the war, and will now be hastened, when, garaging his aircraft at the London Aerodrome as a convenient starting-point, an aerial traveller will tour regularly by air, using his flying machine as he would a motor. Already, dotted about England, are aerodromes he may use as halting-points on his flight, and at which he can house his machine and secure the attention of mechanics; and the number of these grounds should grow rapidly in the future.

In the aeroplane for the tourist, for the man who buys a machine and flies for his own pleasure, it is necessary to combine comfort and safety. As regards comfort, though much remains to be done in the perfection of detail, the occupants of a machine are now more studied than they were in the pioneer days. Then a pilot sat out on a crude seat, exposed fully to the rush of wind as a machine moved through the air. Now he is placed within a covered-in hull, a screen to protect him from the wind. From this stage, as was the case with the motor-car, rapid progress should be made in a provision of comfort.

When touring by air under favourable conditions, there should be no more risk with an aircraft than with a motor-car. One of the most frequent causes of accident, as we have shown, has been the structural weakness of a machine. Now, with the experience of the war on which to draw, and with many clever brains focussed on the development of the industry, this risk may be regarded as almost non-existent; as negligible a factor as it is possible to make it, remembering that aircraft, like other mechanism, have to be built by human hands.

Another risk, that of engine failure, may, as we have explained, be eliminated by the use of more than one motor. In the application of such systems there is still much to be learned; but the obstacles are not insuperable. One advantage that can be offered the aerial tourist, reckoning him as a pilot of no more than average skill, who needs all the aid that science can give him, is that he can obtain a machine which, owing to its automatic stability, requires merely to be taken into the air and brought to earth again, and which will practically fly itself, once it is aloft.

One of the needs with a touring machine, to which makers must devote their attention, is that it should be able to leave the ground quickly in its ascent, and so permit its pilot to rise even from a small starting ground. And it is equally necessary that, on occasion, a machine should be able to alight safely, and at a slow speed, in quite a small field. An aviator who had given up aviation temporarily, after a long spell of cross-country flying, was asked one day when he was going to fly again. "I shan't do so," he said, "till I can buy a machine with which I can alight in my own garden."

Already there are craft which, provided high speeds are not expected of them, and they are given ample plane-surface, will alight at quite a moderate pace; but in the future, by the use of machines which have the power of increasing or reducing their wing-surfaces while in flight, it should be possible to descend in a space no larger, say, than a garden. In the construction of variable-surface machines, technical problems need to be faced which are unusually difficult. The theory with such craft is that their sustaining planes, either by a telescopic system, or by some process of reefing, are built so that they can be expanded or contracted at the will of the pilot. Thus in rising, when a machine is required to ascend with a minimum run forward across the ground, a large area of lifting surface would be exposed; and at the moment of alighting, also, when it was desired that a machine should make its contact with the ground at the slowest possible speed, a maximum of plane surface would be employed. But when aloft, and in full flight, the pilot would be able if he so desired to reduce the area of his lifting surface, and so increase materially his speed. With a machine of this type, when perfected, it should be possible to rise quickly, and descend slowly, and yet at the same time, when well aloft, attain a high speed with moderate engine-power.

The commercial possibilities of aviation are vast and far-reaching: not for nothing, after centuries of striving, have men conquered the air. The aeroplane is destined, by the facilities it offers for communication between nations, to play a vital part in the growth of civilisation. The construction and perfection of a commercial aeroplane, a machine which can be used for the transport of passengers, mails, and goods, represents largely a question of time and of money. Technical problems still need to be solved. But none of them are insurmountable. All should be overcome by an expenditure of money and in the process of time--granted of course that research is directed upon the right lines. A sufficient amount of money for experimental work, which in aviation is very costly, was one of the prime difficulties before the war. Capitalists were chary of aviation; they had no faith in it. Now, after the work aircraft have done in war, and with the need to provide the world with air fleets, the industry need live no longer from hand to mouth. There should be funds available for experiments with commercial-type aeroplanes.

As to the factor of time, this depends largely on the facilities that are obtained by the industry--apart from its work on naval and military craft--for test work with other machines. But in five years' time, granted progress continues on the lines now promised, we should have a service of passenger aeroplanes, each carrying fifty or more people, flying daily between London, the Midlands, and the North; while in ten years' time it should be possible to cross the Atlantic, from London to New York, by means of a regular service of aerial craft.

The commercial aeroplane, even when perfected, would not be likely to compete successfully with other means of transit unless it could offer the advantages of a greater speed. Here, indeed, in the speeds they will attain, lies the future of aircraft. The air will be our highway because, in the air, speeds will be reached that are impossible on land or sea. As civilisation extends--this is of course a truism--there grows with it a need for speedier travel; and we have seen land and sea transit striving to meet this demand. But both have reached, or are rapidly reaching, a limit of speed--a limit imposed by the need to carry their passengers and goods on a remunerative basis. On the sea, by burning excessive quantities of coal, it is possible to add a few knots to the speed of a great liner. But then the problem becomes one of profit and loss; while with trains--so nearly under existing conditions have they reached a limit of speed--that a difficulty is experienced, even on long runs, and under favourable circumstances, in saving a minute here and there. It is not of course to be assumed, when the spur of a greater necessity comes, that land and sea transit will fail altogether to increase their existing speeds. There is the mono-rail system of land traction, electrically propelled, which has yet to be tested in a practical way; while on the sea, perhaps, under pressure of competition, and with an increasing demand for greater speeds, it may be possible to adapt with advantage, even on large craft, some principle of the hydroplane.

But by way of the air, granted even a speeding-up on land and sea, should go the high-speed traffic of the future. By a greater efficiency in lifting surfaces and by reductions in the resistance a craft offers to its own passage through the air; by the provision of systems which will permit a pilot to reduce plane-area when his machine has gained altitude and he desires a maximum speed; by the equipping of craft with motors developing thousands of horse-power for a very low weight--by such means, and by a general improvement in design, it should be possible, eventually, to attain flying speeds of 150, 200, and even 250 miles an hour. From London to New York by air liner, in less than twenty hours; such, for instance, should be an attainment of the future.

It seems probable, in the development of the commercial aeroplane, we shall have machines for touring and for pleasure flights--craft not of large size but in which efforts are made to obtain a greater reliability and comfort. Then it appears likely that aircraft may reach a practical use as carriers of mails and of light express goods; first of all in localities, and under conditions, which favour specially an aerial transit. And from this phase we should move to the passenger-carrying craft; to the days when we shall be able to spend a week-end in New York, as readily as it has been the habit to do in Paris; when we shall be able to reach any part of the world in a journey by air lasting, say, a week or ten days. Then, as a recompense for the lives that have been lost, and for a conquest that has been so dearly won, the world will enter upon an age of aerial transit--the age when frontiers and seas will act as barriers no longer, when journeys that now last weeks will be reduced to days, and those of days to hours; when first of all Europe, and then the world, will be linked by airway.

THE END.